AF388987

Quercus Genetics: Insights into the Past, Present, and Future of Oaks

Quercus Genetics: Insights into the Past, Present, and Future of Oaks

Editors

Mary V. Ashley
Janet R. Backs

MDPI • Basel • Beijing • Wuhan • Barcelona • Belgrade • Manchester • Tokyo • Cluj • Tianjin

Editors
Mary V. Ashley   Janet R. Backs
University of Illinois at Chicago   University of Illinois at Chicago
USA   USA

Editorial Office
MDPI
St. Alban-Anlage 66
4052 Basel, Switzerland

This is a reprint of articles from the Special Issue published online in the open access journal *Forests* (ISSN 1999-4907) (available at: https://www.mdpi.com/journal/forests/special_issues/Quercus_Genetics).

For citation purposes, cite each article independently as indicated on the article page online and as indicated below:

LastName, A.A.; LastName, B.B.; LastName, C.C. Article Title. *Journal Name* **Year**, *Volume Number*, Page Range.

ISBN 978-3-0365-2878-6 (Hbk)
ISBN 978-3-0365-2879-3 (PDF)

Contents

About the Editors

Mary V. Ashley received her doctorate in Biology from the University of California, San Diego, in 1986 and is Professor of Biological Science at the University of Illinois, Chicago. She holds adjunct appointments at the Field Museum and the Chicago Botanic Garden. Her laboratory uses molecular markers to study mating systems and population genetics of both plants and animals. She has authored more than 100 peer reviewed publications and mentored 30 graduate students in her laboratory. Her lab was the first to use microsatellite markers combined with paternity assignment to study pollination patterns in plants, and she provided the first reports of the extent of long-distance pollen movement in wind-pollinated trees. Dr. Ashley has conducted research on several species of threatened *Quercus* in North America. She teaches ecology and evolution at an urban public university with one of the most diverse student bodies in the U.S.A. Her work on promoting women and minorities in STEM has been supported by the National Science Foundation, including a recent grant to increase the success of Latinx students in STEM.

Janet R. Backs is currently a research assistant professor in the Ashley Lab at the University of Illinois, Chicago, where she was awarded a PhD in Biological Sciences in 2014. Her dissertation is entitled "Population structure and gene flow in two rare, isolated *Quercus* species: *Q. hinckleyi* and *Q. pacifica*". She also has master's degrees in Ibero-American Studies (University of Wisconsin, Madison, 1971), Chemistry (Northeastern Illinois University, 1985), and Natural Resources/Environmental Sciences (University of Illinois at Urbana-Champaign, 2008). Prior to her doctoral work, she was a partner in Sobek Consulting, Inc., specializing in database design and support. She has worked extensively with DB2, an IBM database designed for large-scale applications, on multiple platforms as a database administrator and database analyst. She has published articles on population genetics of threatened and endemic oak species, genetic diversity of ex situ oak collections, landscape genetics, oak hybridization, and urban greening, and has presented her research results at the Texas Plant Conservation Conference, the 8th International Oak Society Conference, Botany 2012, and the 8th California Islands Symposium.

Editorial

Quercus Genetics: Insights into the Past, Present, and Future of Oaks

Janet R. Backs * and Mary V. Ashley

Department of Biological Sciences, University of Illinois at Chicago, Chicago, IL 60607, USA; ashley@uic.edu
* Correspondence: jbacks@uic.edu

Citation: Backs, J.R.; Ashley, M.V. Quercus Genetics: Insights into the Past, Present, and Future of Oaks. *Forests* **2021**, *12*, 1628. https://doi.org/10.3390/f12121628

Received: 18 November 2021
Accepted: 22 November 2021
Published: 24 November 2021

The genus *Quercus* comprises over 400 species found across the Northern Hemisphere. Oaks provide important ecosystem services to a multitude of species, including humans, and have shown remarkable resilience and an ability to adapt to variations in climate and geography over the past 56 million years. The tools of genetics and genomics have been important in understanding *Quercus* evolutionary history and distributions, as well as current species delineations, population structure, conservation needs, and the genetic mechanisms that have allowed oaks to adapt to challenging environments. As climate change accelerates, oaks will need to adapt at an even greater pace. Genetics/genomics helps conservationists understand the adaptive mechanisms of oak, and will be valuable for planning management strategies.

In this Special Issue of *Forests*, we have gathered articles describing a range of genetic methods and how they are being applied to the genus *Quercus*. They include investigations into oak evolutionary history, population diversity, gene flow, effects of human disturbance, gaps in genetic research of threated species, and epigenetic modification effects on phenotype variation.

Rapid climate change and anthropogenic disturbances challenge many oak species. The Red List of Oaks 2020 estimates that 41% of oaks are 'species of conservation concern'. A review article identifies threatened *Quercus* species and geographical areas that have gaps in genetic research and delineates genetic methods that would address those gaps [1]. A population study of *Q. ruber* in Lithuania using chloroplast DNA provides insight into postglacial migration of the species and raises the hypothesis that human logging over the last few centuries has reduced existing genetic diversity [2]. A review and synthesis of recent work using RAD-seq-based phylogenies reconstructs the timing and biogeography of North American oak diversification, clarifying relationships and providing new insight into evolutionary radiations [3]. One study used reduced-representation bisulfite sequencing to examine the relationship between genomic and epigenomic variation and observed phenotypic traits. The findings indicate that epigenetic methylation may be a valid marker of phenotype variation, but more research needs to be conducted to determine if it drives variation [4]. Two articles describe studies into genetic variation in and among species, with an emphasis on conservation genetics. One study examines a uniquely adapted desert oak with a disjunct distribution using genetic, morphometric and environmental datasets. The authors present conservation and taxonomic implications of their findings [5]. A second paper examines the genetic diversity of three threatened oaks, comparing levels of diversity to that of common oaks and testing for correlations with range size, population size, and abiotic variations. They also characterize genetic diversity in ex situ collections and demonstrate a need for more extensive geographic sampling [6]. Two papers look at gene flow in oaks. One reviews recent genomic research into oaks' environment-driven adaptive divergence while maintaining species integrity given rates of gene flow, introgression, and hybridization propensity [7]. The second paper reviews studies in *Quercus* pollination using microsatellites and paternity analysis. The paper looks at what has been learned, what questions remain, and how these findings can inform forest management in a rapidly changing world [8].

Funding: This research received no external funding.

Conflicts of Interest: The authors declare no conflict of interest.

References

1. Backs, J.R.; Ashley, M.V. Quercus conservation genetics and genomics: Past, present, and future. *Forests* **2021**, *12*, 882. [CrossRef]
2. Danusevičius, D.; Baliuckas, V.; Buchovska, J.; Kembrytė, R. Geographical structuring of *Quercus robur* (L.) chloroplast DNA haplotypes in Lithuania: Recolonization, adaptation, or overexploitation effects? *Forests* **2021**, *12*, 831. [CrossRef]
3. Manos, P.S.; Hipp, A.L. An updated infrageneric classification of the North American oaks (*Quercus* subgenus *Quercus*): Review of the contribution of phylogenomic data to biogeography and species diversity. *Forests* **2021**, *12*, 786. [CrossRef]
4. Browne, L.; MacDonald, B.; Fitz-Gibbon, S.; Wright, J.W.; Sork, V.L. Genome-wide variation in DNA methylation predicts variation in leaf traits in an ecosystem-foundational oak species. *Forests* **2021**, *12*, 569. [CrossRef]
5. Zumwalde, B.A.; McCauley, R.A.; Fullinwider, I.J.; Ducket, D.; Spence, E.S.; Hoban, S. Genetic, morphological, and environmental differentiation of an arid-adapted oak with a disjunct distribution. *Forests* **2021**, *12*, 465. [CrossRef]
6. Spence, E.S.; Fant, J.; Gailing, O.; Griffith, M.P.; Havens, K.; Hipp, A.L.; Kadav, P.; Kramer, A.; Thompson, P.; Toppila, R.; et al. Comparing genetic diversity in three threatened oaks. *Forests* **2021**, *12*, 561. [CrossRef]
7. Lazic, D.; Hipp, A.L.; Carlson, J.E.; Gailing, O. Use of genomic resources to assess adaptive divergence and introgression in oaks. *Forests* **2021**, *12*, 690. [CrossRef]
8. Ashley, M.V. Answers blowing in the wind: A quarter century of genetic studies of pollination in oaks. *Forests* **2021**, *12*, 575. [CrossRef]

 *forests*

Review

Quercus Conservation Genetics and Genomics: Past, Present, and Future

Janet R. Backs * and Mary V. Ashley

Department of Biological Sciences, University of Illinois at Chicago, Chicago, IL 60607, USA; ashley@uic.edu
* Correspondence: jbacks@uic.edu

Abstract: *Quercus* species (oaks) have been an integral part of the landscape in the northern hemisphere for millions of years. Their ability to adapt and spread across different environments and their contributions to many ecosystem services is well documented. Human activity has placed many oak species in peril by eliminating or adversely modifying habitats through exploitative land usage and by practices that have exacerbated climate change. The goal of this review is to compile a list of oak species of conservation concern, evaluate the genetic data that is available for these species, and to highlight the gaps that exist. We compiled a list of 124 Oaks of Concern based on the Red List of Oaks 2020 and the Conservation Gap Analysis for Native U.S. Oaks and their evaluations of each species. Of these, 57% have been the subject of some genetic analysis, but for most threatened species (72%), the only genetic analysis was done as part of a phylogenetic study. While nearly half (49%) of published genetic studies involved population genetic analysis, only 16 species of concern (13%) have been the subject of these studies. This is a critical gap considering that analysis of intraspecific genetic variability and genetic structure are essential for designing conservation management strategies. We review the published population genetic studies to highlight their application to conservation. Finally, we discuss future directions in *Quercus* conservation genetics and genomics.

Keywords: conservation genetics; conservation genomics; *Quercus*; Oaks of Concern; endangered oaks

Citation: Backs, J.R.; Ashley, M.V. *Quercus* Conservation Genetics and Genomics: Past, Present, and Future. *Forests* **2021**, *12*, 882. https://doi.org/10.3390/f12070882

Academic Editors: Dušan Gömöry and Timothy A. Martin

Received: 1 May 2021
Accepted: 30 June 2021
Published: 6 July 2021

Publisher's Note: MDPI stays neutral with regard to jurisdictional claims in published maps and institutional affiliations.

1. Introduction

Oaks have evolved and adapted over the past 56 million years [1]. Their success has been attributed to high genetic diversity, rapid migration and adaptability, and their propensity for hybridization and introgression [2]. While the ecosystem services oaks provide support a multitude of species, humans in particular have interacted and benefited directly from oaks over many millennia. Acorns of *Quercus ithaburensis* and *Q. caliprinos*, identified by charcoal analysis, have been associated with early humans from 65,000–48,000 years ago and were likely included in their diet [3]. Oak wood, bark, leaves, and roots, as well as acorns, are part of traditional medicine in many parts of the world, and continue to be used as medicinal remedies [4–7]. Oaks figure in human folklore and culture [8] and have been perceived as sacred in numerous human societies [9]. Because of their abundance and high biomass they sequester carbon and so contribute to climate regulation [10]. Genomic studies are now illuminating the genetic basis behind humans' representation of oaks as symbolic of 'longevity, cohesiveness, and robustness' [11]. While this paper focuses on the species that have been adversely impacted by humans, the long relationship between humans and oaks has also had a positive effect on many species [2]. For example, fire regimes that have been used over human history to control undergrowth and enhance hunting areas have benefited oaks through reduced competition with understory vegetation and more shade-tolerant trees in the open areas maintained by such fires [12–14]. However, we are currently living in a time when many oak species are in danger.

The exact number of species in the genus *Quercus* is still open to clarification, as new species continue to be discovered in oak hot-spots such as Mexico/Central America and

China/Southeast Asia. Recent estimates are that approximately 430–435 differentiated species exist [2,15,16]. Oaks occur across the Northern Hemisphere from the equator to boreal regions and thrive in elevations from sea level to 4000 m on various soil types from alkaline to acidic. Species richness is especially high in North America and Asia, where oaks have adapted (and speciated) in response to varying ecological niches [2]. Some oaks, such as *Q. hinckleyi*, are as small as one meter at maturity and grow as clumps of long-lived clones. More familiar to most are large trees that dominate the landscape and live for hundreds of years, such as *Quercus macrocarpa*, *Q. petraea* and *Q. robur*. They can also be extremely rare and critically threatened (*Q. hinckleyi* [17,18]) or abundant with wide-spread distributions, like the other species just mentioned.

Effective conservation management benefits from genetic data to clarify species' identity and adaptations and to provide information on intraspecific diversity and population structure. Unfortunately, missing or incomplete genetic information limits comprehensive planning for many threatened oak species. A survey of conservation actions conducted as part of the Conservation Gap Analysis of Native U.S. Oaks [19] (hereafter Gap Analysis) found that genetic research was one of the least reported efforts, highlighting the need for genetic investigations for many oak species. Specific genetic gaps are seen for phylogenetics/taxonomy to clarify evolutionary significant units (ESU) and for population genetics to address diversity, gene flow, and hybridization/introgression. These findings motivated our effort here to evaluate the state of *Quercus* conservation genetics.

2. Developing a List of Oaks of Concern

For our review of the status of oak conservation genetics, we combine findings from The Red List of Oaks 2020 [16] (hereafter Red List) and the Gap Analysis to create a list of global species of concern. The Red List assessed 430 *Quercus* species. Species were evaluated based on current and projected population sizes, geographic range/endemism, population decline, and fragmentation. While the majority of oak species are not threatened, the study found that 41% are 'species of conservation concern' with 112 falling into the IUCN Threatened Categories: Critically Endangered (CR), Endangered (EN), or Vulnerable (VU) (see p. 9 and Appendix A in [16]) for descriptions and criteria of the IUCN categories. We use this list of species in our review. An additional 105 species are categorized as Near Threatened (NT) or Data Deficient (DD). If DD species are included in the analysis' calculations, the report estimates that globally 31% of oaks are in danger of extinction [16]; we did not include these additional 105 species in our literature review since few have been the subject of genetic studies.

The Red List identifies the U.S. as one of the areas with the highest number of threatened oaks, so we incorporated additional information from a detailed report on U.S. oaks, the Gap Analysis [19]. The Gap Analysis found that 31% of native U.S. oak species are 'species of concern' based on an assessment of data reported in The Red List of Oaks 2017 [20], the NatureServe conservation rankings [21], the USDA Forest Service risk assessment of vulnerability to climate change [22], and a survey of ex situ collections conducted as part of the gap report itself. Evaluation focused on risks of extinction, susceptibility to the effects of climate change, and presence of species in ex situ collections and considered both current and near-term threats. In addition to the criteria used in the Red List, the Gap Analysis looked at regeneration/recruitment and genetic variation/species integrity. Based on these criteria, it scored each species' level of vulnerability. Results show 28 U.S. oak species are 'species of concern', including 12 species not included in the Red List that we added to our list of 'Oaks of Concern'. Our combined list of threatened species is shown in Table 1 and contains 124 species.

Table 1. Species of Concern and genetic research focus.

Species of Concern [Cited Research]	Conservation Classification	Country Distribution	Number of Ex Situ Collections	*Quercus* Section	Citation Focus
Quercus acerifolia [1]	EN	US	44	*Lobatae*	PT
Quercus acutifolia [1,23]	VU	BZ, GT, HN, MX	34	*Lobatae*	PT
Quercus afares [1,24–27]	VU	DZ, TN	18	*Cerris*	PT, PG
Quercus ajoensis [1,28]	VU	MX, US	5	*Quercus*	PT
Quercus albicaulis	CR	CN	0	*Cyclobalanopsis*	
Quercus arbutifolia [1,29,30]	EN	CN, VN	0	*Cyclobalanopsis*	PT, PG
Quercus argyrotricha	CR	N	4	*Cyclobalanopsis*	
Quercus arkansana [1]	VU	US	48	*Lobatae*	PT
Quercus asymmetrica	EN	CN, VN	0	*Cyclobalanopsis*	
Quercus austrina [1,28]	VU	US	25	*Quercus*	PT
Quercus austrocochinchinensis [1,30–34]	VU	CN, LA, TH, VN	0	*Cyclobalanopsis*	PT, PG, G
Quercus bambusifolia [35,36]	EN	CN, HK, VN	4	*Cyclobalanopsis*	PG
Quercus baniensis [33,34]	CR	VN	0	*Cyclobalanopsis*	PT
Quercus baolamensis [34]	CR	VN	0	*Cyclobalanopsis*	PT
Quercus bawanglingensis [37]	CR	CN	0	*Ilex*	PT, G
Quercus bidoupensis [34]	CR	VN	0	*Cyclobalanopsis*	PT
Quercus blaoensis [33,34]	CR	VN	0	*Cyclobalanopsis*	PT
Quercus boyntonii [1,28,38,39]	CR	US	21	*Quercus*	PT, PG, CON
Quercus braianensis [33,34]	VU	LA, VN	0	*Cyclobalanopsis*	PT
Quercus brandegeei [28,40]	EN	MX	9	*Virentes*	PT
Quercus cambodiensis [34]	CR	KH	0	*Cyclobalanopsis*	PT
Quercus camusiae [33,34]	CR	VN	0	*Cyclobalanopsis*	PT
Quercus carmenensis	EN	MX, US	3	*Quercus*	
Quercus cedrosensis [25,41]	VU	MX, US	2	*Protobalanus*	PT
Quercus chapmanii [1,28]	* LC	US	8	*Quercus*	PT
Quercus chrysotricha	EN	MY	0	*Cyclobalanopsis*	
Quercus costaricensis [1]	VU	CR, HN, PA	2	*Lobatae*	PT
Quercus cualensis	EN	MX	1	*Lobatae*	
Quercus cupreata	EN	MX	4	*Lobatae*	
Quercus daimingshanensis (damingshanensis) [1,30]	EN	CN	0	*Cyclobalanopsis*	PT
Quercus dankiaensis	CR	VN	0	*Cyclobalanopsis*	
Quercus delgadoana [1]	EN	MX	9	*Lobatae*	PT
Quercus delicatula	EN	CN	0	*Cyclobalanopsis*	
Quercus devia	EN	MX	0	*Lobatae*	
Quercus dilacerata [34]	CR	VN	0	*Cyclobalanopsis*	PT
Quercus dinghuensis	CR	CN	1	*Cyclobalanopsis*	
Quercus disciformis [33]	EN	CN	2	*Cyclobalanopsis*	PT
Quercus diversifolia [1]	EN	MX	3	*Quercus*	PT
Quercus donnaiensis [33,34]	CR	VN	0	*Cyclobalanopsis*	PT
Quercus dumosa [1,28,42–46]	EN	MX, US	30	*Quercus*	PT, PG
Quercus edithiae [33,47,48]	EN	CN, HK, VN	0	*Cyclobalanopsis*	PT
Quercus engelmannii [1,28,41,44,49–51]	EN	MX, US	36	*Quercus*	PT, PG
Quercus fimbriata	CR	CN	0	*Ilex*	

Table 1. *Cont.*

Species of Concern [Cited Research]	Conservation Classification	Country Distribution	Number of Ex Situ Collections	*Quercus* Section	Citation Focus
Quercus flocculenta	EN	MX	3	*Lobatae*	
Quercus furfuracea	VU	MX	5	*Lobatae*	
Quercus gaharuensis	VU	ID, MY	0	*Cyclobalanopsis*	
Quercus galeanensis	EN	MX	8	*Lobatae*	
Quercus georgiana [1,39,52]	EN	US	55	*Lobatae*	PT, PG
Quercus graciliformis	CR	MX, US	21	*Lobatae*	
Quercus gulielmi-treleasei [23]	VU	CR, PA	2	*Lobatae*	PT
Quercus havardii [1,28,53]	EN	US	19	*Quercus*	PT, PG
Quercus hinckleyi [1,18,54,55]	CR	MX, US	12	*Quercus*	PT, PG, CON
Quercus hintonii [23]	EN	MX	3	*Lobatae*	PT
Quercus hintoniorum [23]	VU	MX	6	*Lobatae*	PT
Quercus hirtifolia	EN	MX	7	*Lobatae*	
Quercus honbaensis [34]	CR	VN	0	*Cyclobalanopsis*	PT
Quercus hondae [56]	VU	JP	2	*Cyclobalanopsis*	PT
Quercus inopina [1]	* LC	US	5	*Lobatae*	PT
Quercus insignis [1,28,57]	EN	BZ, CR, GT, HN, MX, NI, PA	27	*Quercus*	PT
Quercus kerangasensis	VU	BN, ID, MY	0	*Cyclobalanopsis*	
Quercus kinabaluensis	EN	MY	0	*Cyclobalanopsis*	
Quercus kingiana [58]	EN	CN, LA, MM, TH	0	*Ilex*	PT
Quercus kiukiangensis [30]	EN	CN	4	*Cyclobalanopsis*	PT
Quercus kotschyana [1,59]	EN	LB	0	*Quercus*	PT
Quercus kouangsiensis [1]	EN	CN	0	*Cyclobalanopsis*	PT
Quercus laceyi [1,28]	* LC	MX, US	16	*Quercus*	PT
Quercus lenticellata	EN	TH	0	*Cyclobalanopsis*	
Quercus liboensis	EN	CN	2	*Cyclobalanopsis*	
Quercus litseoides	VU	CN, HK	1	*Cyclobalanopsis*	
Quercus lobata [1,28,41,44,49,60–86]	* NT	US	41	*Quercus*	PT, PG, G, CON, GENOM
Quercus lobbii	EN	BD, CN, IN	0	*Cyclobalanopsis*	
Quercus lodicosa	EN	CN, IN, MM	0	*Ilex*	
Quercus look [1,59]	EN	LB, SY	16	*Cerris*	PT
Quercus lungmaiensis	CR	CN	1	*Cyclobalanopsis*	
Quercus macdougallii	EN	MX	0	*Quercus*	
Quercus marlipoensis	CR	CN	1	*Ilex*	
Quercus meavei	VU	MX	1	*Lobatae*	
Quercus merrillii	VU	ID, MY, PH	0	*Cyclobalanopsis*	
Quercus miquihuanensis [23]	EN	MX	12	*Lobatae*	PT
Quercus monnula	CR	CN	0	*Quercus*	
Quercus motuoensis	CR	CN	0	*Cyclobalanopsis*	
Quercus mulleri [87]	CR	MX	0	*Lobatae*	PG
Quercus nivea	EN	MY	0	*Cyclobalanopsis*	
Quercus nixoniana	EN	MX	0	*Lobatae*	
Quercus obconicus	EN	CN	0	*Cyclobalanopsis*	

Table 1. *Cont.*

Species of Concern [Cited Research]	Conservation Classification	Country Distribution	Number of Ex Situ Collections	*Quercus* Section	Citation Focus
Quercus oglethorpensis [1,28,39]	EN	US	47	*Quercus*	PT, PG
Quercus pacifica [1,28,43–46]	EN	US	22	*Quercus*	PT, PG
Quercus palmeri [1,28,41]	* NT	MX, US	18	*Protobalanus*	PT
Quercus parvula [1,88–90]	* NT	US	15	*Lobatae*	PT, PG
Quercus percoriacea	EN	MY	0	*Cyclobalanopsis*	
Quercus petelotii	EN	VN	0	*Cyclobalanopsis*	
Quercus phanera [1]	EN	CN	1	*Cyclobalanopsis*	PT
Quercus pinbianensis	CR	CN	0	*Cyclobalanopsis*	
Quercus pontica [1,25,28,61]	EN	GE, TR	91	*Ponticae*	PT
Quercus pseudosetulosa [58]	CR	CN	0	*Ilex*	PT
Quercus pseudoverticillata	CR	MY	0	*Cyclobalanopsis*	
Quercus pumila [23]	* LC	US	15	*Lobatae*	PT
Quercus quangtriensis [33]	VU	CN, LA, MM, TH, VN	0	*Cyclobalanopsis*	PT
Quercus radiata [1,91]	EN	MX	0	*Lobatae*	PT
Quercus ramsbottomii	EN	MM, TH	0	*Cyclobalanopsis*	
Quercus robusta	* DD	US	2	*Lobatae*	
Quercus rubramenta	VU	MX	0	*Lobatae*	
Quercus runcinatifolia	EN	MX	1	*Lobatae*	
Quercus rupestris	EN	VN	0	*Cyclobalanopsis*	
Quercus sadleriana [1,28,61]	* NT	US	14	*Ponticae*	PT
Quercus sagrana (sagraeana) [28]	EN	CU	1	*Virentes*	PT
Quercus semiserratoides	CR	CN	2	*Cyclobalanopsis*	
Quercus sichourensis [30,48,92]	CR	CN	1	*Cyclobalanopsis*	PT, G
Quercus similis [1,28]	* LC	US	2	*Quercus*	PT
Quercus steenisii	EN	ID	0	*Cyclobalanopsis*	
Quercus tardifolia	* DD	MX, US	0	*Lobatae*	
Quercus thomsoniana	CR	BD, BT, IN	0	*Cyclobalanopsis*	
Quercus tiaoloshanica [93]	EN	CN	0	*Cyclobalanopsis*	CON
Quercus tomentella [1,28,41,61,94–97]	EN	MX, US	33	*Protobalanus*	PT, PG
Quercus tomentosinervis	CR	CN	0	*Cyclobalanopsis*	
Quercus toumeyi [1,28]	* DD	MX, US	3	*Quercus*	PT
Quercus treubiana	VU	ID, MY	0	*Cyclobalanopsis*	
Quercus trungkhanhensis [33]	CR	VN	0	*Ilex*	PT
Quercus tuitensis	VU	MX	0	*Lobatae*	
Quercus tungmaiensis [58,98]	EN	CN, IN	3	*Ilex*	PT, G
Quercus utilis [1,48,58]	EN	CN	2	*Ilex*	PT
Quercus vicentensis	VU	MX, SV	1	*Quercus*	
Quercus xanthotricha	EN	CN, LA	0	*Cyclobalanopsis*	
Quercus xuanlienensis [33]	CR	VN	0	*Cyclobalanopsis*	PT

Conservation classification: CR—Critically Endangered, EN—Endangered, VU—Vulnerable, * NT—Near Threatened, * LC—Least Concern, * DD—Data Deficient (* Gap Analysis). (see [16] for descriptions and criteria of IUCN categories). Country distribution: BD—Bangladesh, BT—Bhutan, BZ—Belize, CN—China, CR—Costa Rica, CU—Cuba, DZ—Algeria, GE—Georgia, GT—Guatemala, HK—Hong Kong, HN—Honduras, ID—Indonesia, IN—India, JP—Japan, KH—Cambodia, LA—Lao People's Dem. Republic, LB—Lebanon, MM—Myanmar, MX—Mexico, MY—Malaysia, NI—Nicaragua, PA—Panama, PH—Philippines, SV—El Salvador, SY—Syrian Arab Republic, TH—Thailand, TN—Tunisia, TR—Turkey, US—United States, VN—Viet Nam. Citation focus: PT—phylogeny/taxonomy, PG—population genetics, CON—conservation, G—genome assembly, GENOM—genomic methods.

The countries with the highest numbers of threatened oak species are China with 36, Mexico with 32, and the United States with 28. Not surprisingly these are the three countries with the highest oak species richness. Other regions of concern are Viet Nam with 20 threatened oak species and Malaysia with nine. Chinese, Mexican, and Vietnamese oaks are mainly threatened by loss of habitat due to logging, agriculture, and urbanization, while in the United States, climate change and invasive species are the major concerns [16].

3. Genetic Research on Oaks of Concern

We used citations in the Red List, the Gap Analysis, Google Scholar Searches (filtering on species name and the word 'genetics' for a date range of 2000 to present), and citations contained within these studies, to identify genetic studies that have been conducted on the species on our list, with the caveat that the results represent a 'point in time' and that new research is constantly being added, may not yet be published, or was not found by this search protocol. Citations to genetic studies are included in Table 1.

We found a total of 78 references that included analysis of one or more species on the list. We classified each of the cited papers by the main focus of the research: phylogeny/taxonomy (PT), population genetics (PG), conservation (CON), genome assembly (G), and genomic methods (GENOM). Of the 124 species, 71 (57%) had one or more published research papers involving a genetic study. *Quercus* sections with the highest numbers of listed species are *Cyclobalanopsis*, *Lobatae*, and *Quercus*. Of the Oaks of Concern in each of these sections, 24 of 55 (44%) of *Cyclobalanopsis*, 15 of 30 (50%) of *Lobatae*, and 16 of 21 (76%) of *Quercus* taxa have cited genetic research.

Of the published genetic papers, 24 of 78 (31%) are phylogeny/taxonomy related (PT) with a number dealing with macroevolution of the *Quercus* genus (such as [1,28]), while others focus on phylogenies within *Quercus* sections (such as [23,40,41,61]), and still others look at regional phylogenies (such as [30,33,34,46–48,56,58,59,88,91]). It is notable that for 51 of the 71 (72%) species with cited works, the *only* genetic research was phylogeny/taxonomy related (see Table 1). The rise of genomic analysis in phylogenetics over the last few years (phylogenomics [99]) has provided better tools for clarifying enigmatic relationships within *Quercus*. In particular, with the application of RAD-seq methods, it has been possible to use tens of thousands of genetic markers to get good phylogenetic signal and provide insights into the evolutionary diversification of the *Quercus* genus [1,2]. Phylogenetic research is important in delineating conservation units (ESUs), a critical component in establishing a starting point for conservation planning. Phylogenetics/phylogenomics has been used to answer many questions about Oaks of Concern, such as confirming species' identity and addressing introgression. Additionally, many of these papers include other pertinent data relating to conservation, for example, to hybridization and biogeography. While elucidating the evolutionary history of oaks is certainly an important endeavor, direct implications of phylogenetic reconstruction for conservation management are limited. Most phylogenetic studies include only one or a very few representatives of each species, so provide little insight into many issues most pertinent to conservation management.

4. Population Genetics for Oaks of Concern

We found that a critical area of conservation research, population genetics studies (PG), are missing for the vast majority of Oaks of Concern. Population genetics studies assess intraspecific genetic diversity and population structure and form the foundation of the field of conservation genetics. While we found 39 of 78 (50%) papers focused on population genetic (PG) questions, only 16 different species were investigated in these papers (Table 2), leaving 87% of the species with no information on intraspecific diversity. Five papers examined specific conservation (CON) questions, such as ex situ conservation [100], habitat destruction [93], and genetics as input to conservation planning in response to climate change [73,76,85]. Five dealt with genomic methods (GENOM) such as epigenetics [79,80,86], ecological niche modeling [84], and landscape genomics [81].

Table 2. Population genetics studies involving Oaks of Concern.

Species of Concern	Citation	Focus of Study	Method Used
Q. afares	Mir et al. [24]	hybridization	nuclear allozymes, chloroplast markers
Q. arbutifolia	Xu et al. [29]	genetic diversity	chloroplast (cpDNA), nuclear (ITS) DNA sequences
Q. austrocochinchinensis	An et al. [31]	introgression	AFLP markers, nu-SSRs
Q. bambusifolia	Zeng et al. [35,36]	inbreeding, genetic diversity, population structure	nu-SSRs
Q. boyntonii	Spence et al. [39]	fragmentation, ex situ collections, inbreeding, heterozygosity	nu-SSRs and EST-SRRs
Q. dumosa	Backs et al. [43]	introgression	nu-SSR
	Burge et al. [42]	gene flow/environmental gradients	RAD-seq
	Ortego et al. [45]	genetic differentiation, population structure	nu-SSR
Q. engelmannii	Oney-Birol et al. [51]	hybridization/introgression	RNA-seq
	Ortego et al. [50]	ecological niche modeling	nu-SSR
	Riordan et al. [49]	responses to geography and climate	nu-SSR
Q. georgiana	Kadav [52]	genetic diversity, population structure	EST-SSRs
	Spence et al. [39]	fragmentation, ex situ collections, inbreeding, heterozygosity	nu-SSRs and EST-SRRs
Q. havardii	Zumwalde et al. [53]	genetic diversity, population structure	nu-SSRs
Q. hinckleyi	Backs et al. [18]	genetic diversity, population structure	nu-SSRs
	Backs et al. [55]	hybridization	nu-SSRs
	Backs et al. [54]	genetic diversity, in situ/ex situ	nu-SSRs
Q. lobata	Abraham et al. [68]	hybridization	nu-SSRs
	Ashley et al. [65]	landscape genetics, population structure	nu-SSRs
	Browne at al. [66]	adaptational lag/temperature	genome-wide sequencing
	Craft et al. [69]	hybridization	nu-SSRs
	Dutech et al. [67]	gene flow, genetic diversity, population structure	nu-SSRs
	Gharehaghaji et al. [70]	gene flow	nu-SSRs
	Grivet et al. [73]	gene flow	nu-SSRs
	Gugger et al. [60]	sequence variation/climate gradients	whole-transcriptome sequencing (mRNA-Seq)
	Mead et al. [64]	ecophysiological traits/gene expression	RNA-seq
	Pluess et al. [74]	gene flow	nu-SSRs
	Scofield et al. [75]	gene flow	nu-SSRs
	Sork et al. [77]	gene flow, pollen movement	allozymes, nu-SSRs
	Sork et al. [78]	gene flow, population structure	nu-SSRs
	Sork et al. [63]	gene flow, environmental gradients	chloroplast and nuclear microsatellite
	Sork et al. [46]	hybridization, introgression	nu-SSRs, RADseq-based sequences

Table 2. *Cont.*

Species of Concern	Citation	Focus of Study	Method Used
Q. mulleri	Pingarroni et al. [87]	genetic diversity	nu-SSRs
Q. oglethorpensis	Spence et al. [39]	fragmentation, ex situ collections, inbreeding, heterozygosity	nu-SSRs and EST-SRRs
Q. pacifica	Backs et al. [43]	gene flow, population structure, introgression	nu-SSRs
	Ortego et al. [45]	evolutionary history, demographics	nu-SSRs, cpSSRs
Q. parvula	Dodd et al. [90]	species differentiation	AFLP genetic markers
	Kashani et al. [89]	genetic differentiation, introgression	AFLP genetic markers
Q. tomentella	Ashley et al. [94,95]	genetic variation, structure	nu-SSRs
	Ashley et al. [95]	landscape and conservation genetics	nu-SSRs
	Ashley et al. [96]	genetic variation, population structure	nu-SSRs

While it is discouraging that only 16 Oaks of Concern have population genetics (PG) related citations (Table 2), these studies provide important examples of the work that has been done, as well as highlighting the somewhat limited breadth of the population genetics research for endangered oaks to date. Of those species that have been studied using population genetic approaches, twelve are in Mexico or the U.S., three in China or Southeast Asia, and one in North Africa. China has the largest number of threatened species on our Oaks of Concern list (36), but only a handful of species have been studied (Table 1) and only three have been the subject of population genetic research (Table 2). These three species are members of the *Cyclobalanopsis* section occurring in Southeast Asia. *Quercus arbutifolia* is an Endangered species found in the mountain cloud forests of southern China and Viet Nam. Xu et al. [29], used chloroplast (cpDNA) and nuclear (ITS) DNA sequences to examine *Q. arbutifolia*'s genetic diversity, phylogeographic structure, and evolutionary history. The authors acknowledge the highly threatened status of this species, but highlight how their findings on genetic diversity, which they found to be unexpectedly high, and population dynamics are critical to developing effective conservation plans. *Quercus austrocochinchinensis* is a Vulnerable species found in China, The Lao People's Democratic Republic, Thailand, and Viet Nam. It is referenced in six genetic research papers, but only one population genetics paper. Possible hybridization between *Q. austrocochinchinensis* and a sympatric species, *Q. kerrii*, was investigated using AFLP markers and nu-SSRs, providing information for long-term conservation and restoration of the tropical ravine rainforest environment in the Indo-China area [31]. *Quercus bambusifolia* is an Endangered species found in China, Hong Kong, and Viet Nam. Using population genetics analyses of nu-SSR data, Zeng et al. [35,36] examined inbreeding, genetic diversity, and population structure, and provide data applicable to restoration of severely fragmented tropical landscapes.

The only North African species on our list of Oaks of Concern is *Q. afares*, the African Oak, a Vulnerable species with a limited distribution in the coastal mountains of Algeria and Tunisia. With genetic analysis using nuclear allozymes and chloroplast markers, Mir et al. [24] confirmed its identity as a stable hybrid of two sympatric but phylogenetically distant species, *Q. suber* and *Q. canariensis*. *Q. afarensis* combines traits of these two species, suggesting one or more hybridization events.

Quercus mulleri (section *Lobatae*) is a microendemic oak found in the Sierra Sur de Oaxaca of Mexico. In the first report on this species since it was identified 60 years ago, a population genetics study (PG) examined genetic diversity and population structure using nu-SSRs with the goal of providing information to enhance conservation strategies [87].

The western United Sates, particularly California, has been the focus of numerous studies of oak population genetics, including several studies of Oaks of Concern. *Quercus dumosa* is an endangered oak (section *Quercus*) found in Baja California, Mexico, and

California. Three population genetics studies (PG) examined genetic exchange between this species and its close relatives. One study focused on the influence of environmental gradients using RAD-seq [42], another examined introgression and species' integrity related to neighboring species using nu-SSR [43], and a third examined genetic differentiation and population structure to examine evolutionary history of sympatric species using nu-SSR [45]. *Quercus engelmanii* is also an endangered species distributed in Baja California, Mexico and southern California, US. Population genetic studies have examined responses to geography and climate [49], ecological niche modeling [50], and hybridization/introgression using RNA-seq [51].

Valley oak, *Q. lobata*, a California endemic listed as Near Threatened, is by far the most thoroughly investigated species on our list. Seven papers report on phylogeny/taxonomy [1,28,44,61,62,71,83], one paper reports on genome assembly [62], and five (including one in this special edition [86]) investigate new genomic methods 79–81,84,86]. A number of population genetic papers have also focused on *Q. lobata*. Several studies have investigated gene flow, hybridization, and population structure [46,65,67–70,72,74,75,77,78]. One study used whole-transcriptome sequencing (mRNA-Seq) to investigate sequence variation with climate gradients [60] and another looked at geographic patterns of genetic variation in relation to climate change using chloroplast and nu-SSRs [63]. Drought response was measured using ecophysiological traits and gene expression (RNA-seq) [64]. A common garden experiment and genome-wide sequencing were used to examine adaptational lag to temperature, with potential application to identifying genotypes preadapted to future climate change conditions [66]. Valley oak provides a model of how different genetic approaches can be used to investigate the ecological and evolutionary genetics of a threatened tree species, predict future trends, and assist in developing strategies to manage the risks a species is facing.

Two species of oaks on our list are endemic to the California Channel Islands, *Q. pacifica* and *Q. tomentella*. *Quercus pacifica* is an Endangered species, and researchers have investigated gene flow, population structure, and relationships to two mainland oaks using nu-SSRs [43] and evolutionary history and demographics using nu-SSRs and cpSSRs [45]. *Quercus tomentella*, also listed as Endangered, is a member of the small section *Protobalanus*. Population genetics papers cover genetic variation and population structure [94], landscape and conservation genetics [95] and genetic variation and population structure [96] all using nu-SSRs. Another species, *Quercus parvula* is found on Santa Cruz Island and in the California Coast Ranges and is classified as Near Threatened. Two population genetics papers report on genetic differentiation and introgression [89] and species differentiation [90], both using AFLP genetic markers.

Quercus hinckleyi is a Critically Endangered species with an extremely limited distribution in Texas, USA. Population genetic studies have examined clonality, diversity and population structure [18], hybridization [55], and genetic diversity assessment of in situ and ex situ populations [54]. Important for conservation, Backs et al. [18] reported a high level of clonality at some sites, with the number of genetically unique individuals being substantially lower than previously assumed from population counts.

The other Oaks of Concern that have had population genetic studies are primarily in the Eastern United States. One paper in this Special Issue examines genetic diversity and population structure in the Endangered *Q. havardii* [53]. Another paper in this Special Issue conducted population assessments of three Oaks of Concern, *Q. georgiana*, *Q. oglethorpensis*, and *Q. boyntonii*, and reports that these species have lower genetic diversity than more abundant oaks [39]. Another paper covers genetic diversity and population structure of *Q. georgiana* using EST-SSRs [52].

One question recently explored for several North America species is how well the genetic diversity of the species is captured in ex situ collections. Oak seeds (acorns) are not candidates for seed banks; they lose viability when desiccated. Desiccation is part of the standard protocol of conventional seed-banking [101–104]. Conservation of living oaks in ex situ collections can be constrained by space limitations, long generation times, and their proclivity to hybridize [54]. A recent paper by Backs et al. [54] reported that ex situ collections of *Q. hinckleyi* were likely sampled from only one of the remaining in situ genetic clusters and missed much of the in situ diversity. The study of *Quercus georgiana*, *Q. oglethorpensis*, and *Q. boyntonii* mentioned above reports that while common alleles are well preserved in ex situ collection, low frequency and rare alleles are not [39].

For many other species on our Oaks of Concern list, there are immediate questions that can be addressed through genetic and/or genomic analysis, such as confirming taxa, identifying clones, examining levels of diversity, and measuring gene flow [105]. *Quercus tardifolia*, is a good example. It is described as Data Deficient, requiring both field research and taxonomic clarification and lacking demographic data and diversity information [19]. Other species of concern are not so lacking in information, but still have genetic gaps in their conservation portfolios. *Quercus robusta* needs research to distinguish spontaneous hybrids from historic hybrids that have evolved into true species [106]. For *Q. acerifolia*, there is a need to identify genetic structure of populations [107] and for *Q. carmenensis* there is a need to verify species integrity and/or levels of introgression [108].

To summarize, our survey of genetic studies shows that while some Oaks of Concern have benefited from population genetic research, most are lacking basic conservation-focused data. This need can be addressed through population genetic analysis looking at species integrity, intraspecific diversity, population structure, gene flow, hybridization levels, and diversity capture in ex situ collections. Only with this information can comprehensive conservation strategies be developed.

5. Future Directions

Some exciting new genetic methods that have application to conservation questions can be characterized as 'genomic research'. The ability to look across an entire oak genome or use 'reduced representation' genome sampling has been made possible in recent years by advances in DNA sequencing as well as through better 'big data' manipulation due to increased computing power, storage capabilities, and robust analytic applications. A sampling of genomic research that has been directed toward oak conservation includes phylogenomics [83], epigenetics [80], QTL (quantitative trait loci) [109], and landscape genomics [110]. While phylogenomics provides a broad evolutionary picture of oaks and helps define ESU's, epigenetics, QTL, and landscape genomics have the potential to investigate the genomic basis of adaptive traits and apply this knowledge to developing conservation strategies for Oaks of Concern. These new genomic research methods will provide information on plant and species adaptive responses, data needed for flexible conservation efforts that may include plant migration and/or reintroduction.

Epigenetics is the study of heritable phenotypic changes in an organism that do not involve alterations in the DNA code itself. It is emerging as an important field of research for understanding plant adaptability and plasticity and for identifying the 'ecological background' of individuals [111]. The shortfall in oak-related research in these areas is exemplified by a survey of epigenetic research which found of approximately 20,000 epigenetic studies published in 2019, only 3% of the papers were plant-related, and of those only 5% focused on forest species [112]. Of tree-related papers, only a handful reference *Quercus* species, for example, [80,113,114].

Epigenetic modification can be created by biotic or abiotic environmental stresses, stochastic "epigenetic mutations" [111,115] or natural processes such as hybridization [116]. They can be reset when a stress is relieved or may result in heritable epigenetic marks that can be passed on as 'molecular memory' persisting through several subsequent generations and potentially becoming evolutionarily viable [116–118]. Oaks as long lived organisms have the time to enhance their epigenetic responses through a number of stressful events before passing along these responses through their germlines [119]. Some work has been done with oaks and epigenetics [114,120] including a paper in this Special Issue that examines experimental DNA methylation using the Near Threatened *Q. lobata* [86]. This is an area of conservation interest that should be explored both for planning conservation strategies and basic research into underlying adaptive mechanisms.

Current data processing capabilities have also made it possible to search genome-wide for QTL (quantitative trait loci) [109]. QTL mapping seeks to identify the relationship between various genomic locations and a set of quantitative traits, leading to a chromosomal location and ultimately to identification of gene(s) with the final goal of looking at gene expression. Among other things, this will lead to a better understanding of genetic mechanisms of variation and adaptation [121]. Results can then be applied to adjust conservation measures in response to rapid change, for example, by identifying the genetic adaptability potential of individuals to be used in assisted migration or reintroduction [122,123].

Landscape genomics examines the relationship and interaction between adaptive genetic loci on genomes and landscape variations across which natural populations exist [110]. It seeks to identify the aspects of the environment that affect genetic variation and how that variation in turn affects adaptation [124,125]. It is a valuable tool in understanding oaks' responses to environmental stresses and evaluating alleles that occur under certain climate and habitat conditions. These correlations will aid in conservation planning for plant migration or restoration by identifying populations or individuals that are currently responding favorably to conditions that are anticipated in the climatic future [81,126,127].

These emerging genomic tools as well as more traditional population genetic analyses can and should provide vital input to developing effective oak conservation strategies. Our review highlighted important gaps in our knowledge of many species that are or may soon be facing extinction. For geneticists, there is much work and many opportunities to address conservation needs of the Oaks of Concern.

Funding: This research received no external funding.

Institutional Review Board Statement: Not applicable.

Informed Consent Statement: Not applicable.

Data Availability Statement: Not applicable.

Conflicts of Interest: The authors declare no conflict of interest.

References

1. Hipp, A.L.; Manos, P.S.; Hahn, M.; Avishai, M.; Bodénès, C.; Cavender-Bares, J.; Crowl, A.A.; Deng, M.; Denk, T.; Fitz-Gibbon, S.; et al. Genomic landscape of the global oak phylogeny. *New Phytol.* **2020**, *226*, 1198–1212. [CrossRef]
2. Kremer, A.; Hipp, A.L. Oaks: An evolutionary success story. *New Phytol.* **2020**, *226*, 987–1011. [CrossRef]
3. Lev, E.; Kislev, M.E.; Bar-Yosef, O. Mousterian vegetal food in Kebara Cave, Mt. Carmel. *J. Archaeol. Sci.* **2005**, *32*, 475–484. [CrossRef]
4. Morales, D. Oak trees (*Quercus* spp.) as a source of extracts with biological activities: A narrative review. *Trends Food Sci. Technol.* **2021**, *109*, 116–125. [CrossRef]
5. Lev, E.; Amar, Z. Ethnopharmacological survey of traditional drugs sold in Israel at the end of the 20th century. *J. Ethnopharmacol.* **2000**, *72*, 191–205. [CrossRef]
6. Guha, B.; Arman, M.; Islam, M.N.; Tareq, S.M.; Rahman, M.M.; Sakib, S.A.; Mutsuddy, R.; Tareq, A.M.; Emran, T.B.; Alqahtani, A.M. Unveiling pharmacological studies provide new insights on *Mangifera longipes* and *Quercus gomeziana*. *Saudi J. Biol. Sci.* **2021**, *28*, 183–190. [CrossRef] [PubMed]
7. Xu, J.; Fu, C.; Li, T.; Xia, X.; Zhang, H.; Wang, X.; Zhao, Y. Protective effect of acorn (*Quercus liaotungensis* Koidz) on streptozotocin-damaged MIN6 cells and type 2 diabetic rats via p38 MAPK/Nrf2/HO-1 pathway. *J. Ethnopharmacol.* **2021**, *266*, 113444. [CrossRef] [PubMed]
8. Logan, W.B. *Oak: The Frame of Civilization*; WW Norton & Company: New York, NY, USA, 2005.
9. Altman, N. *Sacred Trees*; Sierra Club Books: San Francisco, CA, USA, 1994.
10. Cavender-Bares, J. Diversity, distribution and ecosystem services of the North American oaks. *Int. Oaks* **2016**, *27*, 37–48.
11. Leroy, T.; Plomion, C.; Kremer, A. Oak symbolism in the light of genomics. *New Phytol.* **2020**, *226*, 1012–1017. [CrossRef]
12. Keeley, J.E. Native American impacts on fire regimes of the California coastal ranges. *J. Biogeogr.* **2002**, *29*, 303–320. [CrossRef]
13. Barrett, R.D.; Schluter, D. Adaptation from standing genetic variation. *Trends Ecol. Evol.* **2008**, *23*, 38–44. [CrossRef]
14. Dey, D.C.; Guyette, R.P. Anthropogenic fire history and red oak forests in south-central Ontario. *For. Chron.* **2000**, *76*, 339–347. [CrossRef]

15. Denk, T.; Grimm, G.W.; Manos, P.S.; Deng, M.; Hipp, A.L. An updated infrageneric classification of the oaks: Review of previous taxonomic schemes and synthesis of evolutionary patterns. In *Oaks Physiological Ecology. Exploring the Functional Diversity of Genus Quercus L.*; Springer: Cham, Switzerland, 2017; pp. 13–38.

16. Carrero, C.; Jerome, D.; Beckman, E.; Byrne, A.; Coombes, A.J.; Deng, M.; González-Rodríguez, A.; Hoang, V.S.; Khoo, E.; Nguyen, N.; et al. *The Red List of Oaks*; The Morton Arboretum: Lisle, IL, USA, 2020.

17. Muller, C.H. The signigicance of vegetative reproduction in *Quercus. Madroño* **1951**, *11*, 129–137.

18. Backs, J.R.; Terry, M.; Klein, M.; Ashley, M. V Genetic analysis of a rare isolated species: A tough little West Texas oak, *Quercus hinckleyi* CH Mull. *J. Torrey Bot. Soc.* **2015**, *142*, 302–313. [CrossRef]

19. Beckman, E.; Meyer, A.; Denvir, A.; Gill, D.; Man, G.; Pivorunas, D.; Shaw, K.; Westwood, M. *Conservation Gap Analysis of Native U.S. Oaks.*; The Morton Arboretum: Lisle, IL, USA, 2019.

20. Jerome, D.; Beckman, E.; Kenny, L.; Wenzell, K.; Kua, C.; Westwood, M. *The Red List of US Oaks*; Morton Arboretum: Lisle, IL, USA, 2017.

21. Frances, A. *NatureServe and Native Plant Conservation in North America*; Ormsby, A., Leopold, S., Eds.; NatureServe: Arlington, VA, USA, 2017.

22. Potter, K.M.; Crane, B.S.; Hargrove, W.W. A United States national prioritization framework for tree species vulnerability to climate change. *New For.* **2017**, *48*, 275–300. [CrossRef]

23. Vázquez, M.L. Molecular evolution of the internal transcribed spacers in red oaks (*Quercus* sect. *Lobatae*). *Comput. Biol. Chem.* **2019**, *83*, 107117. [CrossRef]

24. Mir, C.; Toumi, L.; Jarne, P.; Sarda, V.; Di Giusto, F.; Lumaret, R. Endemic North African *Quercus afares* Pomel originates from hybridisation between two genetically very distant oak species (*Q. suber* L. and *Q. canariensis* Willd.): Evidence from nuclear and cytoplasmic markers. *Heredity* **2006**, *96*, 175–184. [CrossRef] [PubMed]

25. Denk, T.; Grimm, G.W. The oaks of western Eurasia: Traditional classifications and evidence from two nuclear markers. *Taxon* **2010**, *59*, 351–366. [CrossRef]

26. Sakka, H.; Baraket, G.; Abdessemad, A.; Tounsi, K.; Ksontini, M.; Salhi-Hannachi, A. Molecular phylogeny and genetic diversity of Tunisian *Quercus* species using chloroplast DNA CAPS markers. *Biochem. Syst. Ecol.* **2015**, *60*, 258–265. [CrossRef]

27. Simeone, M.C.; Cardoni, S.; Piredda, R.; Imperatori, F.; Avishai, M.; Grimm, G.W.; Denk, T. Comparative systematics and phylogeography of *Quercus* Section *Cerris* in western Eurasia: Inferences from plastid and nuclear DNA variation. *PeerJ* **2018**, *6*, e5793. [CrossRef]

28. McVay, J.D.; Hipp, A.L.; Manos, P.S. A genetic legacy of introgression confounds phylogeny and biogeography in oaks. *Proc. R. Soc. Lond. B Biol. Sci.* **2017**, *284*, 20170300. [CrossRef] [PubMed]

29. Xu, J.; Jiang, X.-L.; Deng, M.; Westwood, M.; Song, Y.-G.; Zheng, S.-S. Conservation genetics of rare trees restricted to subtropical montane cloud forests in southern China: A case study from *Quercus arbutifolia* (Fagaceae). *Tree Genet. Genomes* **2016**, *12*, 1. [CrossRef]

30. Deng, M.; Jiang, X.L.; Hipp, A.L.; Manos, P.S.; Hahn, M. Phylogeny and biogeography of East Asian evergreen oaks (*Quercus* section *Cyclobalanopsis*; Fagaceae): Insights into the Cenozoic history of evergreen broad-leaved forests in subtropical Asia. *Mol. Phylogenet. Evol.* **2018**, *119*, 170–181. [CrossRef] [PubMed]

31. An, M.; Deng, M.; Zheng, S.-S.; Jiang, X.-L.; Song, Y.-G. Introgression Threatens the Genetic Diversity of *Quercus austrocochinchinensis* (Fagaceae), an Endangered Oak: A Case Inferred by Molecular Markers. *Front. Plant Sci.* **2017**, *8*, 229. [CrossRef] [PubMed]

32. An, M.; Deng, M.; Zheng, S.-S.; Song, Y.-G. De novo transcriptome assembly and development of SSR markers of oaks *Quercus austrocochinchinensis* and *Q. kerrii* (Fagaceae). *Tree Genet. Genomes* **2016**, *12*, 1. [CrossRef]

33. Binh, H.T.; Ngoc, N.V.; Bon, T.N.; Tagane, S.; Suyama, Y.; Yahara, T. A new species and two new records of Quercus (Fagaceae) from northern Vietnam. *PhytoKeys* **2018**, *92*, 1–15. [CrossRef] [PubMed]

34. Binh, H.T.; Ngoc, N.V.; Tagane, S.; Toyama, H.; Mase, K.; Mitsuyuki, C.; Strijk, J.S.; Suyama, Y.; Yahara, T. A taxonomic study of *Quercus langbianensis* complex based on morphology and DNA barcodes of classic and next generation sequences. *PhytoKeys* **2018**, *95*, 37–70. [CrossRef]

35. Zeng, X.; Fischer, G.A. Using multiple seedlots in restoration planting enhances genetic diversity compared to natural regeneration in fragmented tropical forests. *For. Ecol. Manag.* **2021**, *482*, 118819. [CrossRef]

36. Zeng, X.; Fischer, G.A. Wind pollination over 70 years reduces the negative genetic effects of severe forest fragmentation in the tropical oak *Quercus bambusifolia*. *Heredity* **2020**, *124*, 156–169. [CrossRef] [PubMed]

37. Liu, X.; Chang, E.-M.; Liu, J.-F.; Huang, Y.-N.; Wang, Y.; Yao, N.; Jiang, Z.-P. Complete chloroplast genome sequence and phylogenetic analysis of *Quercus bawanglingensis* Huang, Li et Xing, a vulnerable oak tree in China. *Forests* **2019**, *10*, 587. [CrossRef]

38. Hoban, S.; Callicrate, T.; Clark, J.; Deans, S.; Dosman, M.; Fant, J.; Gailing, O.; Havens, K.; Hipp, A.L.; Kadav, P.; et al. Taxonomic similarity does not predict necessary sample size for ex situ conservation: A comparison among five genera. *Proc. R. Soc. B* **2020**, *287*, 20200102. [CrossRef]

39. Spence, E.S.; Fant, J.; Gailing, O.; Griffith, M.P.; Havens, K.; Hipp, A.L.; Kadav, P.; Kramer, A.; Thompson, P.; Toppila, R.; et al. Comparing genetic diversity in three threatened oaks. *Forests* **2021**, *12*, 561. [CrossRef]

40. Cavender-Bares, J.; González-Rodríguez, A.; Eaton, D.A.R.; Hipp, A.A.L.; Beulke, A.; Manos, P.S. Phylogeny and biogeography of the American live oaks (*Quercus* subsection *Virentes*): A genomic and population genetics approach. *Mol. Ecol.* **2015**, *24*, 3668–3687. [CrossRef] [PubMed]

41. Manos, P.S.; Doyle, J.J.; Nixon, K.C. Phylogeny, Biogeography, and Processes of Molecular Differentiation in Quercus Subgenus Quercus (Fagaceae). *Mol. Phylogenet. Evol.* **1999**, *12*, 333–349. [CrossRef]

42. Burge, D.O.; Parker, V.T.; Mulligan, M.; Sork, V.L. Influence of a climatic gradient on genetic exchange between two oak species. *Am. J. Bot.* **2019**, *106*, 864–878. [CrossRef] [PubMed]

43. Backs, J.R.; Ashley, M.V. Evolutionary history and gene flow of an endemic island oak: *Quercus pacifica*. *Am. J. Bot.* **2016**, *103*, 2115–2125. [CrossRef] [PubMed]

44. Kim, B.Y.; Wei, X.; Fitz-Gibbon, S.; Lohmueller, K.E.; Ortego, J.; Gugger, P.F.; Sork, V.L. RADseq data reveal ancient, but not pervasive, introgression between Californian tree and scrub oak species (*Quercus* Fagaceae). *Mol. Ecol.* **2018**, *27*, 4556–4571. [CrossRef]

45. Ortego, J.; Noguerales, V.; Gugger, P.F.; Sork, V.L. Evolutionary and demographic history of the Californian scrub white oak species complex: An integrative approach. *Mol. Ecol.* **2015**, *24*, 6188–6208. [CrossRef]

46. Sork, V.L.; Riordan, E.; Gugger, P.F.; Fitz-Gibbon, S.; Wei, X.; Ortego, J. Phylogeny and introgression of California scrub white oaks (*Quercus* section *Quercus*). *Int. Oaks* **2016**, *27*, 61–74.

47. Yang, Y.; Zhu, J.; Feng, L.; Zhou, T.; Bai, G.; Yang, J.; Zhao, G. Plastid genome comparative and phylogenetic analyses of the key genera in Fagaceae: Highlighting the effect of codon composition bias in phylogenetic inference. *Front. Plant Sci.* **2018**, *9*, 82. [CrossRef]

48. Yang, Y.; Zhou, T.; Qian, Z.; Zhao, G. Phylogenetic relationships in Chinese oaks (Fagaceae, *Quercus*): Evidence from plastid genome using low-coverage whole genome sequencing. *Genomics* **2021**, *113*, 1438–1447. [CrossRef] [PubMed]

49. Riordan, E.C.; Gugger, P.F.; Ortego, J.; Smith, C.; Gaddis, K.; Thompson, P.; Sork, V.L. Association of genetic and phenotypic variability with geography and climate in three southern California oaks. *Am. J. Bot.* **2016**, *103*, 73–85. [CrossRef] [PubMed]

50. Ortego, J.; Riordan, E.C.; Gugger, P.F.; Sork, V.L. Influence of environmental heterogeneity on genetic diversity and structure in an endemic southern Californian oak. *Mol. Ecol.* **2012**, *21*, 3210–3223. [CrossRef] [PubMed]

51. Oney-Birol, S.; Fitz-Gibbon, S.; Chen, J.-M.; Gugger, P.F.; Sork, V.L. Assessment of shared alleles in drought-associated candidate genes among southern California white oak species (*Quercus* Quercus). *BMC Genet.* **2018**, *19*, 1. [CrossRef] [PubMed]

52. Kadav, P.D. *Characterization of Genic Microsatellite Markers (EST-SSRSs) in the Endangered Oak Species Quercus georgiana M.A. Curtis*; Michigan Technological University: Houghton, MI, USA, 2017.

53. Zumwalde, B.A.; McCauley, R.A.; Fullinwider, I.J.; Ducket, D.; Spence, E.S.; Hoban, S. Genetic, morphological, and environmental differentiation of an arid-adapted oak with a disjunct distribution. *Forests* **2021**, *12*, 465. [CrossRef]

54. Backs, J.R.; Hoban, S.; Ashley, M.V. Genetic diversity assessment of ex situ collections of endangered *Quercus hinckleyi*. *Int. J. Plant Sci.* **2021**, *182*, 220–228. [CrossRef]

55. Backs, J.R.; Terry, M.; Ashley, M.V. Using genetic analysis to evaluate hybridization as a conservation concern for the threatened species *Quercus hinckleyi* C.H. Muller (Fagaceae). *Int. J. Plant Sci.* **2016**, *177*, 122–131. [CrossRef]

56. Kamiya, K.; Harada, K.; Ogino, K.; Clyde, M.M.; Latiff, A.M. Phylogeny and genetic variation of Fagaceae in tropical montane forests. *Tropics* **2003**, *13*, 119–125. [CrossRef]

57. Rodríguez-Correa, H.; Oyama, K.; Quesada, M.; Fuchs, E.J.; Quezada, M.; Ferrufino, L.; Valencia-Ávalos, S.; Cascante-Marín, A.; González-Rodríguez, A. Complex phylogeographic patterns indicate Central American origin of two widespread Mesoamerican *Quercus* (Fagaceae) species. *Tree Genet. Genomes* **2017**, *13*, 62. [CrossRef]

58. Jiang, X.; Hipp, A.L.; Deng, M.; Su, T.; Zhou, Z.; Yan, M. East Asian origins of European holly oaks (*Quercus* section *Ilex* Loudon) via the Tibet-Himalaya. *J. Biogeogr.* **2019**, *46*, 2203–2214. [CrossRef]

59. Douaihy, B.; Saliba, C.; Stephan, J.; Simeone, M.C.; Cardoni, S.; Farhat, P.; Kharrat, M.B.D. Tracking diversity and evolutionary pathways of Lebanese oak taxa through plastome analyses. *Bot. Lett.* **2020**, *167*, 315–330. [CrossRef]

60. Gugger, P.F.; Cokus, S.J.; Sork, V.L. Association of transcriptome-wide sequence variation with climate gradients in valley oak (*Quercus lobata*). *Tree Genet. Genomes* **2016**, *12*, 15. [CrossRef]

61. Crowl, A.A.; Manos, P.S.; McVay, J.D.; Lemmon, A.R.; Lemmon, E.M.; Hipp, A.L. Uncovering the genomic signature of ancient introgression between white oak lineages (*Quercus*). *New Phytol.* **2020**, *226*, 1158–1170. [CrossRef]

62. Sork, V.L.; Fitz-Gibbon, S.T.; Puiu, D.; Crepeau, M.; Gugger, P.F.; Sherman, R.; Stevens, K.; Langley, C.H.; Pellegrini, M.; Salzberg, S.L. First draft assembly and annotation of the genome of a California endemic oak *Quercus lobata* Née (Fagaceae). *G3 Genes Genomes Genet.* **2016**, *6*, 3485–3495. [CrossRef] [PubMed]

63. Sork, V.L.; Davis, F.W.; Westfall, R.; Flint, A.; Ikegami, M.; Wang, H.; Grivet, D. Gene movement and genetic association with regional climate gradients in California valley oak (*Quercus lobata* Née) in the face of climate change. *Mol. Ecol.* **2010**, *19*, 3806–3823. [CrossRef] [PubMed]

64. Mead, A.; Peñaloza Ramirez, J.; Bartlett, M.K.; Wright, J.W.; Sack, L.; Sork, V.L. Seedling response to water stress in valley oak (*Quercus lobata*) is shaped by different gene networks across populations. *Mol. Ecol.* **2019**, *28*, 5248–5264. [CrossRef]

65. Ashley, M.V.; Abraham, S.T.; Backs, J.R.; Koenig, W.D. Landscape genetics and population structure in Valley Oak (*Quercus lobata* Née). *Am. J. Bot.* **2015**, *102*, 2124–2131. [CrossRef] [PubMed]

66. Browne, L.; Wright, J.W.; Fitz-Gibbon, S.; Gugger, P.F.; Sork, V.L. Adaptational lag to temperature in valley oak (*Quercus lobata*) can be mitigated by genome-informed assisted gene flow. *Proc. Natl. Acad. Sci. USA* **2019**, *116*, 25179–25185. [CrossRef]

67. Dutech, C.; Sork, V.L.; Irwin, A.J.; Smouse, P.E.; Davis, F.W. Gene flow and fine-scale genetic structure in a wind-pollinated tree species, *Quercus lobata* (Fagaceaee). *Am. J. Bot.* **2005**, *92*, 252–261. [CrossRef]

68. Abraham, S.; Zaya, D.; Koenig, W.; Ashley, M. Interspecific and intraspecific pollination patterns of Valley Oak, *Quercus lobata*, in a mixed stand in coastal central California. *Int. J. Plant Sci.* **2011**, *172*, 691–699. [CrossRef]

69. Craft, K.J.; Ashley, M.V.; Koenig, W.D. Limited hybridization between *Quercus lobata* and *Quercus douglasii* (Fagaceae) in a mixed stand in central coastal California. *Am. J. Bot.* **2002**, *89*, 1792–1798. [CrossRef]

70. Gharehaghaji, M.; Minor, E.S.; Ashley, M.V.; Abraham, S.T.; Koenig, W.D. Effects of landscape features on gene flow of valley oaks (*Quercus lobata*). *Plant Ecol.* **2017**, *218*, 487–499. [CrossRef]

71. Grivet, D.; Deguilloux, M.-F.; Petit, R.J.; Sork, V.L. Contrasting patterns of historical colonization in white oaks (*Quercus* spp.) in California and Europe. *Mol. Ecol.* **2006**, *15*, 4085–4093. [CrossRef]

72. Grivet, D.; Robledo-Arnuncio, J.J.; Smouse, P.E.; Sork, V.L. Relative contribution of contemporary pollen and seed dispersal to the effective parental size of seedling population of California valley oak (*Quercus lobata*, Née). *Mol. Ecol.* **2009**, *18*, 3967–3979. [CrossRef]

73. Grivet, D.; Sork, V.L.; Westfall, R.D.; Davis, F.W. Conserving the evolutionary potential of California valley oak (*Quercus lobata* Née): A multivariate genetic approach to conservation planning. *Mol. Ecol.* **2008**, *17*, 139–156. [CrossRef]

74. Pluess, A.R.; Sork, V.L.; Dolan, B.; Davis, F.W.; Grivet, D.; Merg, K.; Papp, J.; Smouse, P.E. Short distance pollen movement in a wind-pollinated tree, *Quercus lobata* (Fagaceae). *For. Ecol. Manag.* **2009**, *258*, 735–744. [CrossRef]

75. Scofield, D.G.; Alfaro, V.R.; Sork, V.L.; Grivet, D.; Martinez, E.; Papp, J.; Pluess, A.R.; Koenig, W.D.; Smouse, P.E. Foraging patterns of acorn woodpeckers (*Melanerpes formicivorus*) on valley oak (*Quercus lobata* Née) in two California oak savanna-woodlands. *Oecologia* **2011**, *166*, 187–196. [CrossRef]

76. Sork, V.L.; Davis, F.W.; Grivet, D. Incorporating Genetic Information into Conservation Planning for California Valley Oak. In Proceedings of the Sixth Symposium on Oak Woodlands: Today's Challenges, Tomorrow's Opportunities, Rohnert Park, CA, USA, 9–12 October 2006; pp. 497–509.

77. Sork, V.L.; Davis, F.W.; Smouse, P.E.; Apsit, V.J.; Dyer, R.J.; Fernandez-M, J.F.; Kuhn, B. Pollen movement in declining populations of California Valley oak, Quercus lobata: Where have all the fathers gone? *Mol. Ecol.* **2002**, *11*, 1657–1668. [CrossRef] [PubMed]

78. Sork, V.L.; Smouse, P.E.; Grivet, D.; Scofield, D.G. Impact of asymmetric male and female gamete dispersal on allelic diversity and spatial genetic structure in valley oak (*Quercus lobata* Née). *Evol. Ecol.* **2015**, *29*, 927–945. [CrossRef]

79. Browne, L.; Mead, A.; Horn, C.; Chang, K.; Celikkol, Z.A.; Henriquez, C.L.; Ma, F.; Beraut, E.; Meyer, R.S.; Sork, V.L. Experimental DNA demethylation associates with changes in growth and gene expression of oak tree seedlings. *G3 Genes Genomes Genet.* **2020**, *10*, 1019–1028. [CrossRef]

80. Gugger, P.F.; Fitz-Gibbon, S.; Pellegrini, M.; Sork, V.L. Species-wide patterns of DNA methylation variation in *Quercus lobata* and its association with climate gradients. *Mol. Ecol.* **2016**, *25*, 1665–1680. [CrossRef] [PubMed]

81. Gugger, P.F.; Fitz-Gibbon, S.T.; Albarrán-Lara, A.; Wright, J.W.; Sork, V.L. Landscape genomics of *Quercus lobata* reveals genes involved in local climate adaptation at multiple spatial scales. *Mol. Ecol.* **2021**, *30*, 406–423. [CrossRef] [PubMed]

82. Gugger, P.F.; Peñaloza-Ramírez, J.M.; Wright, J.W.; Sork, V.L.; Schnitzler, J.-P. Whole-transcriptome response to water stress in a California endemic oak, *Quercus lobata*. *Tree Physiol.* **2016**, *37*, 632. [CrossRef] [PubMed]

83. Fitz-Gibbon, S.; Hipp, A.L.; Pham, K.K.; Manos, P.S.; Sork, V.L.; Jaramillo-Correa, J.P. Phylogenomic inferences from reference-mapped and de novo assembled short-read sequence data using RADseq sequencing of California white oaks (*Quercus* section *Quercus*). *Genome* **2017**, *60*, 743–755. [CrossRef]

84. Gugger, P.F.; Ikegami, M.; Sork, V.L. Influence of late Quaternary climate change on present patterns of genetic variation in valley oak, *Quercus lobata* Née. *Mol. Ecol.* **2013**, *22*, 3598–3612. [CrossRef]

85. Sork, V.L.; Squire, K.; Gugger, P.F.; Steele, S.E.; Levy, E.D.; Eckert, A.J. Landscape genomic analysis of candidate genes for climate adaptation in a California endemic oak, *Quercus lobata*. *Am. J. Bot.* **2016**, *103*, 33–46. [CrossRef]

86. Browne, L.; MacDonald, B.; Fitz-Gibbon, S.; Wright, J.W.; Sork, V.L. Genome-wide variation in DNA methylation predicts variation in leaf traits in an ecosystem-foundational oak species. *Forests* **2021**, *12*, 569. [CrossRef]

87. Pingarroni, A.; Molina-Garay, C.; Rosas-Osorio, C.; Alfonso-Corrado, C.; Clark-Tapia, R.; Monsalvo-Reyes, A.; Campos, J.E. Abundancia y diversidad genética de *Quercus mulleri*, especie microendémica amenazada de Oaxaca. *Madera Bosques* **2020**, *26*. [CrossRef]

88. Hauser, D.A.; Keuter, A.; McVay, J.D.; Hipp, A.L.; Manos, P.S. The evolution and diversification of the red oaks of the California Floristic Province (*Quercus* section Lobatae, series Agrifoliae). *Am. J. Bot.* **2017**, *104*, 1581–1595. [CrossRef]

89. Kashani, N.; Dodd, R.S. Genetic differentiation of two California red oak species, *Quercus parvula* var. *shreveii* and *Q. wislizeni*, based on AFLP genetic markers. In *Proceedings of the Fifth Symposium on Oak Woodlands: Oaks in California's Challenging Landscape*; San Diego, CA, USA, 22–25 October 2001, pp. 417–426.

90. Dodd, R.S.; Kashani, N. Molecular differentiation and diversity among the California red oaks (Fagaceae; *Quercus* section Lobatae). *Theor. Appl. Genet. Theor. Angew. Genet. TAG.* **2003**, *107*, 884–892. [CrossRef]

91. McCauley, R.A.; Cortés-Palomec, A.C.; Oyama, K. Species diversification in a lineage of Mexican red oak (Quercus section Lobatae subsection *Racemiflorae*)—The interplay between distance, habitat, and hybridization. *Tree Genet. Genomes* **2019**, *15*, 1. [CrossRef]

92. Su, H.; Yang, Y.; Ju, M.; Li, H.; Zhao, G. Characterization of the complete plastid genome of Quercus sichourensis. *Conserv. Genet. Resour.* **2019**, *11*, 129–131. [CrossRef]

93. Zheng, J.-W.; An, S.-Q.; Chen, L.; Leng, X.; Wang, Z.-S.; Xiang, H.-J. Effects of logging on the genetic diversity of *Quercus tiaoloshanica* Chun et Ko in a tropical montane forest of Hainan Island, Southern China. *J. Integr. Plant Biol.* **2005**, *47*, 1184–1192. [CrossRef]

94. Ashley, M.V.; Abraham, S.; Kindsvater, L.C.; Knapp, D.; Craft, K. Population structure and genetic variation of Island. In *Oak Ecosystem Restoration Research on Catalina Island, California*; Knapp, D.A., Ed.; Catalina Island Conservancy: Avalon, CA, USA, 2010; pp. 125–135.

95. Ashley, M.V.; Backs, J.R.; Abraham, S.T. Landscape and conservation genetics of the Island Oak, *Quercus tomentella*. *Int. Oaks* **2016**, *27*, 83–90.

96. Ashley, M.V.; Backs, J.R.; Kindsvater, L.; Abraham, S.T. Genetic variation and structure in an endemic island oak, *Quercus tomentella*, and mainland canyon oak, *Quercus chrysolepis*. *Int. J. Plant Sci.* **2018**, *179*, 151–161. [CrossRef]

97. Ortego, J.; Gugger, P.F.; Sork, V.L. Genomic data reveal cryptic lineage diversification and introgression in Californian golden cup oaks (section *Protobalanus*). *New Phytol.* **2018**, *218*, 804–818. [CrossRef]

98. Yang, Y.; Zhang, H.; Ren, T.; Zhao, G. Characterization of the complete plastid genome of Quercus tungmaiensis. *Conserv. Genet. Resour.* **2018**, *10*, 457–460. [CrossRef]

99. Philippe, H.; Delsuc, F.; Brinkmann, H.; Lartillot, N. Phylogenomics. *Annu. Rev. Ecol. Evol. Syst.* **2005**, *36*, 541–562. [CrossRef]

100. Hoban, S. New guidance for ex situ gene conservation: Sampling realistic population systems and accounting for collection attrition. *Biol. Conserv.* **2019**, *235*, 199–208. [CrossRef]

101. Wallace, S.H. Development of an Informational Resource to Inform Global Prioritization of Efforts to Conserve Threatened, Exceptional Plant Taxa. Ph.D. Thesis, University of Delaware, Newark, Delaware, 2015.

102. Wyse, S.V.; Dickie, J.B. Predicting the global incidence of seed desiccation sensitivity. *J. Ecol.* **2017**, *105*, 1082–1093. [CrossRef]

103. Wyse, S.V.; Dickie, J.B.; Willis, K.J. Seed banking not an option for many threatened plants. *Nat. Plants* **2018**, *4*, 848–850. [CrossRef] [PubMed]

104. Pence, V.C. Evaluating costs for the in vitro propagation and preservation of endangered plants. *In Vitro Cell. Dev. Biol. Plant* **2011**, *47*, 176–187. [CrossRef]

105. Van Dyke, F.; Lamb, R.L. Conservation Genetics. In *Conservation Biology: Foundations, Concepts, Applications*; Springer International Publishing: Cham, Switzerland, 2020; pp. 171–210, ISBN 978-3-030-39534-6.

106. Beckman, E.; McNeil-Marshall, A.; Still, S.M.; Meyer, A.; Westwood, M. *Quercus robusta C.H.Müll. Conservation Gap Analysis of Native U.S. Oaks*; The Morton Arboretum: Lisle, IL, USA, 2019; pp. 184–189.

107. Beckman, E.; Baker, B.; Lobdell, M.; Meyer, A.; Westwood, M. *Conservation Gap Analysis of Native U.S. Oaks*; The Morton Arboretum: Lisle, IL, USA, 2019; pp. 50–55.

108. Beckman, E.; Still, S.M.; Meyer, A.; Westwood, M. *Quercus carmenensis C.H.Müll. Conservation Gap Analysis of Native U.S. Oaks*; The Morton Arboretum: Lisle, IL, USA, 2019; pp. 80–85.

109. Borevitz, J.O.; Chory, J. Genomics tools for QTL analysis and gene discovery. *Curr. Opin. Plant Biol.* **2004**, *7*, 132–136. [CrossRef]

110. Storfer, A.; Patton, A.; Fraik, A.K. Navigating the Interface Between Landscape Genetics and Landscape Genomics. *Front. Genet.* **2018**, *9*, 68. [CrossRef]

111. Rey, O.; Eizaguirre, C.; Angers, B.; Baltazar-Soares, M.; Sagonas, K.; Prunier, J.G.; Blanchet, S.; Herrel, A. Linking epigenetics and biological conservation: Towards a conservation epigenetics perspective. *Funct. Ecol.* **2020**, *34*, 414–427. [CrossRef]

112. Amaral, J.; Ribeyre, Z.; Vigneaud, J.; Sow, M.D.; Fichot, R.; Messier, C.; Pinto, G.; Nolet, P.; Maury, S. Advances and promises of epigenetics for forest trees. *Forests* **2020**, *11*, 976. [CrossRef]

113. Rico, L.; Ogaya, R.; Barbeta, A.; Peñuelas, J.; Rennenberg, H. Changes in DNA methylation fingerprint of *Quercus ilex* trees in response to experimental field drought simulating projected climate change. *Plant Biol.* **2014**, *16*, 419–427. [CrossRef] [PubMed]

114. Silva, H.G.; Sobral, R.S.; Magalhães, A.P.; Morais-Cecílio, L.; Costa, M.M.R. Genome-wide identification of epigenetic regulators in *Quercus suber* L. *Int. J. Mol. Sci.* **2020**, *21*, 3783. [CrossRef]

115. Verhoeven, K.J.F.; Preite, V. Epigenetic variation in asexually reproducing organisms. *Evol. Int. J. Org. Evol.* **2013**, *68*, 644–655. [CrossRef]

116. Bräutigam, K.; Vining, K.J.; Lafon-Placette, C.; Fossdal, C.G.; Mirouze, M.; Marcos, J.G.; Fluch, S.; Fraga, M.F.; Guevara, M.Á.; Abarca, D.; et al. Epigenetic regulation of adaptive responses of forest tree species to the environment. *Ecol. Evol.* **2013**, *3*, 399–415. [CrossRef]

117. Bossdorf, O.; Richards, C.L.; Pigliucci, M. Epigenetics for ecologists. *Ecol. Lett.* **2008**, *11*, 106–115. [CrossRef]

118. Chinnusamy, V.; Zhu, J.-K. Epigenetic regulation of stress responses in plants. *Curr. Opin. Plant Biol.* **2009**, *12*, 133–139. [CrossRef]

119. Boyko, A.; Kovalchuk, I. Epigenetic control of plant stress response. *Environ. Mol. Mutagen.* **2008**, *49*, 61–72. [CrossRef] [PubMed]

120. Inácio, V.; Barros, P.M.; Costa, A.; Roussado, C.; Gonçalves, E.; Costa, R.; Graça, J.; Oliveira, M.M.; Morais-Cecílio, L.; Fu, B. Differential DNA Methylation Patterns Are Related to Phellogen Origin and Quality of Quercus suber Cork. *PLoS ONE* **2017**, *12*, e0169018. [CrossRef]

121. Saintagne, C.; Bodénès, C.; Barreneche, T.; Pot, D.; Plomion, C.; Kremer, A. Distribution of genomic regions differentiating oak species assessed by QTL detection. *Heredity* **2004**, *92*, 20–30. [CrossRef]
122. Aitken, S.N.; Whitlock, M.C. Assisted gene flow to facilitate local adaptation to climate change. *Annu. Rev. Ecol. Evol. Syst.* **2013**, *44*, 367–388. [CrossRef]
123. Gonzalez, A.; Ronce, O.; Ferriere, R.; Hochberg, M.E. Evolutionary rescue: An emerging focus at the intersection between ecology and evolution. *Philos. Trans. R. Soc. London* **2013**, *368*, 20120404. [CrossRef] [PubMed]
124. Rellstab, C.; Gugerli, F.; Eckert, A.J.; Hancock, A.M.; Holderegger, R. A practical guide to environmental association analysis in landscape genomics. *Mol. Ecol.* **2015**, *24*, 4348–4370. [CrossRef] [PubMed]
125. Li, Y.; Zhang, X.-X.; Mao, R.-L.; Yang, J.; Miao, C.-Y.; Li, Z.; Qiu, Y.-X. Ten years of landscape genomics: Challenges and opportunities. *Front. Plant Sci.* **2017**, *8*, 2136. [CrossRef] [PubMed]
126. Rellstab, C.; Zoller, S.; Walthert, L.; Lesur, I.; Pluess, A.R.; Graf, R.; Bodénès, C.; Sperisen, C.; Kremer, A.; Gugerli, F. Signatures of local adaptation in candidate genes of oaks (*Quercus* spp.) with respect to present and future climatic conditions. *Mol. Ecol.* **2016**, *25*, 5907–5924. [CrossRef]
127. Martins, K.; Gugger, P.F.; Llanderal-Mendoza, J.; González-Rodríguez, A.; Fitz-Gibbon, S.T.; Zhao, J.-L.; Rodríguez-Correa, H.; Oyama, K.; Sork, V.L. Landscape genomics provides evidence of climate-associated genetic variation in Mexican populations of *Quercus rugosa*. *Evol. Appl.* **2018**, *11*, 1842–1858. [CrossRef] [PubMed]

Article

Geographical Structuring of *Quercus robur* (L.) Chloroplast DNA Haplotypes in Lithuania: Recolonization, Adaptation, or Overexploitation Effects?

Darius Danusevičius [1,*], Virgilijus Baliuckas [1,2], Jurata Buchovska [1,2] and Rūta Kembrytė [1]

[1] Agriculture Academy, Faculty of Forest Sciences and Ecology, Vytautas Magnus University, K. Donelaičio g. 58, LT-44248 Kaunas, Lithuania; virgilijus.baliuckas@lammc.lt (V.B.); jurata.buchovska@vdu.lt (J.B.); ruta.kembryte@vdu.lt (R.K.)

[2] Institute of Forestry, Lithuanian Research Centre for Agriculture and Forestry, Liepų Str. 1, Girionys, LT-53101 Kaunas, Lithuania

* Correspondence: darius.danusevicius@vdu.lt

Abstract: We studied the maternally inherited chloroplast DNA polymorphism at three microsatellite loci of 157 *Quercus robur* trees from 38 native populations in Lithuania. We found high diversity of eight haplotypes from the Balkan lineage A (frequency 0.75) and the "German" subbranch of the Balkan lineage A (freq. 0.12), western and eastern Italian lineages C (freq. 0.05 and 0.06, respectively), and Iberian lineage B (freq. 0.03). The haplotypes were geographically well structured (among population differentiation index PhiPT = 0.30, the *p*-value < 0.001) that is unexpected for such a small territory as Lithuania. We raised a hypothesis on historical overexploitation of oaks by eliminating certain haplotypes in Lithuania, following a drastic felling of oak forests over the last few centuries.

Keywords: chlorotype; pedunculate oak; oak decline; genetic diversity; gene conservation; postglacial migration

Citation: Danusevičius, D.; Baliuckas, V.; Buchovska, J.; Kembrytė, R. Geographical Structuring of *Quercus robur* (L.) Chloroplast DNA Haplotypes in Lithuania: Recolonization, Adaptation, or Overexploitation Effects?. *Forests* **2021**, *12*, 831. https://doi.org/10.3390/f12070831

Academic Editors: Mary Ashley and Janet R. Backs

Received: 3 May 2021
Accepted: 13 June 2021
Published: 24 June 2021

Publisher's Note: MDPI stays neutral with regard to jurisdictional claims in published maps and institutional affiliations.

1. Introduction

Quercus robur (L.) is an important forest key-habitat species, an indicator of the autochthonous forests with immense values for economy, ecosystem services and culture in Europe [1]. In Lithuania alone occupying merely ca. 63 K km^2, there are 74 settlements bearing the name of "oak" [2]. Revealing the evolutionary history of such an ecologically important species would markedly benefit the basic knowledge on genetic structure of forest tree populations and help to efficiently design genetic biodiversity conservation programs [3]. The efficiency here is understood as capturing rare genetic variants rather than the genetic variation common elsewhere. There is a possibility for these rare variants to reside in divergent evolutionary lineages of forest tree populations [4]. Another issue is the decline of health of *Q. robur* forests observed over the recent decades in Europe [5–9]. One possible cause of the oak decline could be depleted genetic diversity due to the centuries of exploitation of oak forests in Europe [5].

A strategy to test the oak decline hypothesis is to study richness and structuring of evolutionary lineages of oaks within a certain region [10]. Presence of structuring of evolutionary lineages, that may be difficult to explain due to natural causes, may be a strong indicator of undesirable human interventions such as overexploitation. The evolutionary origin of plants is commonly determined by studying the genetic variation in maternally inherited chloroplast genomes [11]. In oaks, chloroplast DNA is maternally inherited and is passed over generations less affected by stochastic recombination events as in the nuclear genome [11,12]. Mutations in tree plastid genomes are not frequent and the consensus is that certain maternally inherited DNA lineages (haplotypes) are associated to the glacial refugium populations and the postglacial migration lineages of trees expanding since the LGM at 20,000 years BP in late Pleistocene geological epoch [13,14]. Studies on the fossil pollen abundance in soil sedimental horizons and DNA polymorphism of maternally inherited plastid genomes suggest that genetically distinct zones for postglacial expansion of trees were located in the Iberian and Italian peninsulas, and the Balkans [15,16]. Factors other than the geographic proximity may also affect migration of trees, e.g., river flow direction [17,18]. Based on maternally inherited chloroplast DNA markers the following three main postglacial migration lineages of *Q. robur* were detected in Europe [14,19–21] (Figure 1): Lineage "A" originating from glacial refugium in the Balkans during the migration accumulated mutations leading to several haplotypes, one of which outbranched this lineage in present-day Germany and is often referred to as "German" haplotype; Lineage "B" out of the Iberian peninsula spreading mainly to France and Britain but also reaching as far as the Scandinavian peninsula; Lineage "C" from southern Italy that mutated into western and eastern subbranches, correspondingly spreading towards north-west and north-east of Europe, the latter being common in the western regions of European Russia [21]. Geophysical barriers and corridors as well as proximity to the glacial refugia were the main determinants of the present-day geographical arrangement of the post-glacial migration lineages of *Q. robur*. For instance, the corridor between the Alps and the Carpathian Mountain chains as well as the northward alignment of the Carpathians favored the Balkan refugia of plants for migration northwards over the Italian refugia.

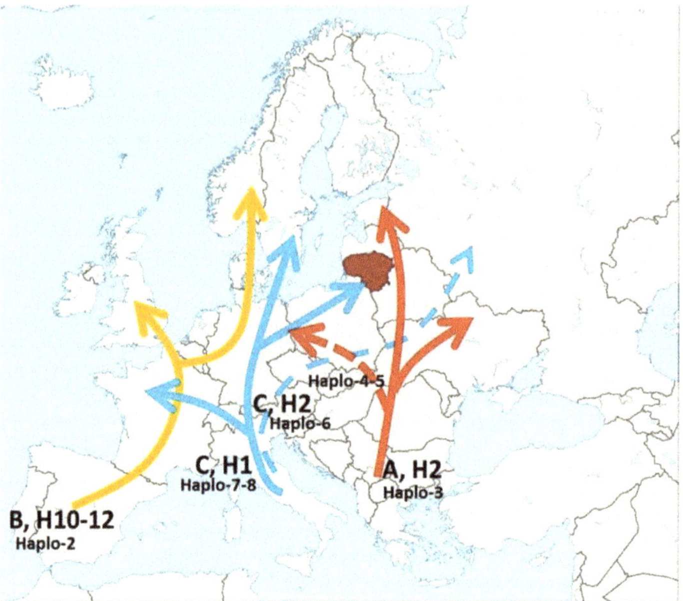

Figure 1. Simplified pathways of the main postglacial migration lineages of *Q. robur* based on [19] (noted by the capital letters A, B, C) and haplotypes as in [14] (noted by capital *H*) and our study (noted by Haplo). Haplo-4–5 indicates the "German" branch of the Balkan lineage A. The location of Lithuania on the map is highlighted.

The Balkan lineage (A) is the geographically closest migration route into the eastern Baltic region [14] and was found to be more frequent in earlier studies on cpDNA markers of oaks in the Baltic region [10,22]. These studies reported high chloroplast DNA haplotype diversity with evidence of geographically non-random distribution raising a hypothesis of local adaptation of certain haplogroups. However, ref. [10] focused on broader eastern Baltic gradient with few populations from Lithuania. Both these studies used the PCR_RFLP DNA marker system which is not precise enough to detect certain haplotypes that can be detected with the chloroplast microsatellite markers [20,21,23].

The radiocarbon dating of the fossil pollen in Europe suggests a fast postglacial tree expansion northward at an average rate of 150 km per century (conifers), 100 km/century (*Corylus, Ulmus, Alnus*) and 40 km/century *(Tilia, Quercus, Fagus Carpinus)* mainly during 14,000 to 12,000 years BP following the warming at the start of the Holocene geological epoch [15,24–26]. The species composition of European forests at certain periods of time depended on (a) the geographical distance from these glacial migration sources (e.g., [14,27]) and (b) once populated, on the long-term temperature fluctuations usually referred to as climate chronozones affecting the species biological adaptability to certain climates [24]. Therefore, each part of Europe has its own time scale for forest species abundance and composition. Based on range wide radiocarbon dated fossil pollen records [28], oaks populated present-day southern France and Germany during the relatively cooler Pre-boreal and Boreal climatic chronozones (10,000–8000 years BP) and reached the eastern Baltic region between 8000 and 6000 years BP during the warm Atlantic climatic chronozone. According to [26], the most abundant oak fossil pollen records north of the latitude 47° N were for 11,000–8000 years BP. However, the oak fossil pollen records in southern Lithuania indicate frequency of oaks up to 1% of the other tree species already during the tundra vegetation period 14,000–13,000 BC [29]. The oldest radiocarbon dated oak fossil wood from mature tree trunks found in the eastern Baltic are of ca. 9000 years BP (fossils conserved at the Baltic seabed or pond peat layer in Birzai, northern Lithuania [30]). Based on fossil pollen, the golden age of oaks in Lithuania was 8000–3000 years BC, with a peak at late Atlantic zone in 6500 years BC [29]. Around 6500 years BC, oaks admixed with other broadleaf trees may have formed several continuous forest tracts and amounted up to 10% of other forest tree species [29]. These large oak dominated forest tracts were located (a) in south-western Lithuania stretching from the central part of Kaunas-Kedainiai via Kalzu Ruda towards Marijampole and southwards, (b) in the western Lithuanian highland (Zemaitija, centered about Rietavas), and (c) in an area from Kaunas stretching eastwards towards Trakai [29]. Since then, following the Subatlantic cooling with expansion of Norway spruce, oak abundance was decreasing gradually down to 5% in the late Subboreal zone (3000 BP) and 1–2% at 500 BP and the present-day.

Strong and dynamic genetic differentiation due to divergent adaptive environments at the range widelevel have led to a pronounced ecotypic structuring within *Q. robur* species *sensu lato* [31–33]. This strong genetic differentiation [34] provided a vast morphological variation for the taxonomists seeking the *sensu stricto* taxonomic categorization [35,36]. Nature, however, seems to complicate the hard work of the taxonomists even more by allowing a spontaneous hybridization between the taxonomic units not only within but also among the species [37,38] and not only for *Quercus* sp. but also for other forest trees [39–41]. Nevertheless, there is a possibility for a long-term retention of the morphology markers that are neutral to selection such as some of the leaf or acorn traits in *Q. robur* [33]. In such case, these neutral marker traits to some degree could be reflected by distinct evolutionary lineages initiated from the divergent glacial refugia [31]. For *Q. robur*, several intraspecific taxonomic structures stand out: *Q. robur* subsp. *robur* (common in central and northern Europe); *Q. robur* subsp. *broteroana* (Schwarz) Camus, *Q. robur* subsp. *estremadurensis* (Schwarz) Camus, common in Spain (both correspond to the Iberian post-glacial migration lineage); *Q. robur* subsp. *Brutia* common in southern Italy (the migration lineages from Italy), *Q. robur* subsp. *slavonica* (Gayer) Matyas, concentrated in Slovenia (the migration lineages from the Balkan Peninsula, [42]), *Q. robur* subsp. *Pedunculiflora* (Asia Minor, Caucasus, Balkans). Based on historical sources, in the 16th century, *Q. robur* stands may have occupied 10–20% of the forests in Lithuania [2]. The paleontological records indicate a lower margin of this range [29]. In any case, since the 18th century in the Russian Empire, a severe felling of oaks for shipbuilding and construction reduced the *Q. robur* stands down to 2–3% of its original range by the 1900s [2]. Therefore, it is likely that humans have contributed to the oak decline phenomena by eroding the genetic diversity when selectively removing oaks of high commercial value and by deforesting areas on rich soils suitable for agriculture, as in south-western Lithuania. Presently, *Q. robur* stands in Lithuania are small and fragmented and occupy 47.3 thousand ha, making up 2.30% of the remaining forests [42]. Commonly, the commercial value of oaks in Lithuania is poor: forked trees, usually losing the apical dominance at the lower half of the stem, dominate (personal observation). There are initiatives at the level of Ministry of Environment to promote and enlarge areas of *Q. robur* in Lithuania. The question of a depleted genetic diversity of *Q. robur* populations in Lithuania is very much relevant, when initiating such large-scale oak restitution initiatives.

The objectives of our study were to assess geographical variation patterns in the maternally inherited chloroplast DNA haplotypes among natural *Q. robur* populations in Lithuania. To supplement the earlier RFLP-based studies on the evolutionary origin of oaks in the Baltics, here we used a dense population sampling, microsatellite genotyping and discussed the main factors that could have influenced the observed structure of oak haplotypes.

2. Material and Methods

We sampled leaves (where accessible) or wood from 157 young trees at 38 locations (populations) in natural stands of *Q. robur* representing all of Lithuania (Figure 2). The young trees were chosen for easy access to leaves. Per location, we selected ca. 5–10 trees spaced more than 30 m apart. By considering the relatively lower variation at the organelle DNA loci, our approach was to capture a more detailed geographical grid at the cost of lower sample size at each grid point. When interpreting the genetic structure, it is easy to visually merge the grid locations into geographically larger units to obtain higher within unit sample size, if needed. The initial aim was to genotype 10 trees per location, however, due to the genotype scoring quality at most of the locations the sample size is lower (Table 1).

Table 1. Location of the populations and the number of the cpSSR haplotypes found in each population. N is the sample size.

Pop	Lat	Long	N	Haplo-1	Haplo-2	Haplo-3	Haplo-4	Haplo-5	Haplo-6	Haplo-7	Haplo-8
				NF	B Iberia H10-12	A Balcan H5-7	A CEU H4	A CEU H4	C ITeast H2	C ITwest H1	C ITwest H1
				141-93-81	141-94-80	142-93-81	142-94-80	142-94-81	143-92-81	143-94-80	143-94-81
Total			157	1	4	117	7	11	10	1	6
ANYK	55.39	25.18	5	0	0	1	0	4	0	0	0
BIRZ	56.17	24.35	5	0	0	5	0	0	0	0	0
DRUS	54.00	24.17	5	0	0	0	2	0	3	0	0
DUBR	54.54	23.48	1	0	0	1	0	0	0	0	0
DZUK	54.06	24.30	1	0	0	0	0	0	1	0	0
IGNA	55.21	26.20	2	0	0	2	0	0	0	0	0
JONI	56.19	23.32	1	0	0	0	1	0	0	0	0
JURB	55.04	22.41	5	0	0	5	0	0	0	0	0
KEDA	55.18	23.66	3	0	0	3	0	0	0	0	0
KRET	55.93	21.14	10	0	0	10	0	0	0	0	0
KUPI	55.48	24.50	2	0	0	1	0	1	0	0	0
KURS	56.03	22.55	2	0	0	1	0	0	0	0	1
MARI	54.43	22.58	5	0	0	5	0	0	0	0	0
MAZE	56.14	22.10	6	0	0	1	1	0	0	0	4
NERI	55.34	21.07	7	0	4	1	2	0	0	0	0
PAKR	55.73	23.56	3	0	0	2	1	0	0	0	0
PRIE	54.35	24.10	6	0	0	6	0	0	0	0	0
RADV	55.38	23.52	7	0	0	7	0	0	0	0	0
RASE	55.16	22.93	6	0	0	6	0	0	0	0	0
ROKI	55.93	25.35	6	1	0	3	0	2	0	0	0

Table 1. *Cont.*

Pop	Lat	Long	N	Haplo-1	Haplo-2	Haplo-3	Haplo-4	Haplo-5	Haplo-6	Haplo-7	Haplo-8
SAKI	54.84	23.01	5	0	0	5	0	0	0	0	0
SALC	54.26	25.14	5	0	0	3	0	0	2	0	0
SIAU	55.85	23.11	3	0	0	3	0	0	0	0	0
SILU	55.37	21.28	6	0	0	5	0	0	0	1	0
SVEN	55.07	26.09	2	0	0	1	0	0	1	0	0
TAUR	55.21	22.28	2	0	0	2	0	0	0	0	0
TELS	55.51	22.00	3	0	0	2	0	0	0	0	1
TYTU	55.33	22.80	9	0	0	7	0	2	0	0	0
UKME	55.10	24.88	8	0	0	8	0	0	0	0	0
UTEN	55.28	25.38	6	0	0	3	0	2	1	0	0
VALK	54.21	24.53	4	0	0	3	0	0	1	0	0
VARE	54.24	24.38	1	0	0	1	0	0	0	0	0
VEIS	54.05	23.39	6	0	0	6	0	0	0	0	0
VILN	54.45	25.13	2	0	0	1	0	0	1	0	0
ZARA	55.48	25.68	7	0	0	7	0	0	0	0	0

The haplotypes detected in our study are coded as Haplo-1 to 8 in the topmost row of the table. The corresponding postglacial migration lineages from [19] are given as capital letters (A, B or C) and codes e.g., H10–12 as in [14] together with location of glacial refugium (CEU is an outbranch of the Balkan lineage distinguished by mutation in central Europe, IT east is eastern Italian lineage, IT west is western Italian lineage). NF means not found in other studies. SSR scores are the cpSSR allele size (in bp).

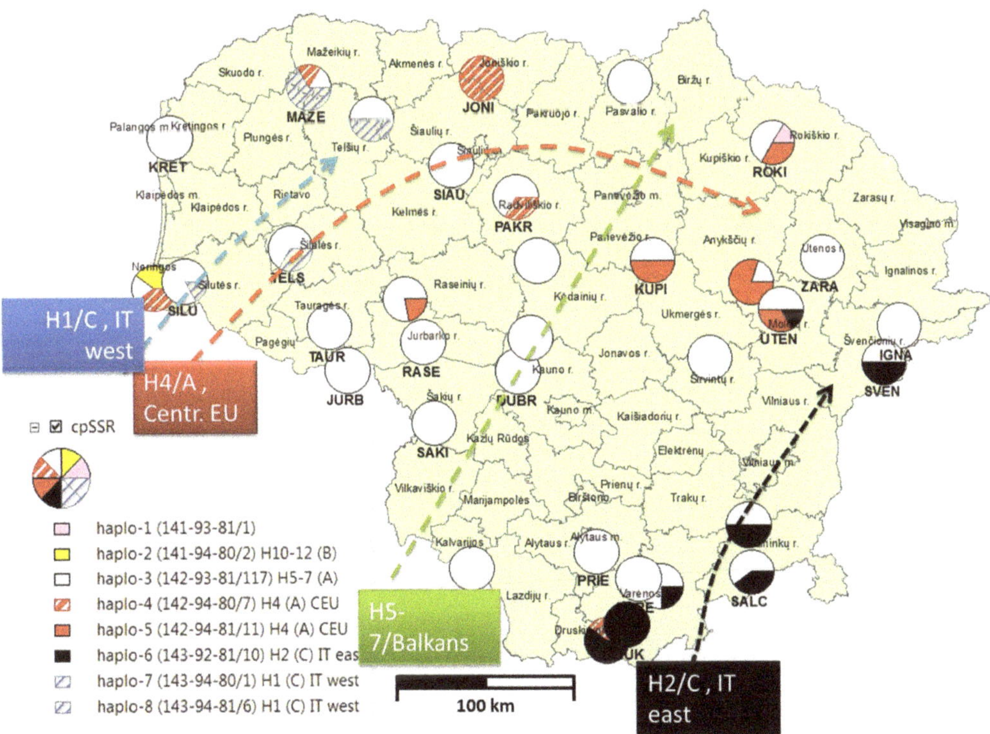

Figure 2. Geographical distribution of *Q. robur* chloroplast DNA haplotypes in Lithuania. The haplotypes are explained in the legend, where the haplotype id, allele scores (both as in our study), number of trees after the slash, the haplotypes as in [14], and the post-glacial migration lineages as in [19] are given. The dashed arrows indicate hypothetical migration routes of the postglacial migration linages identified as in [19] assuming natural migration being the main driver of the present distribution of the haplotypes.

We extracted the DNA from leaves or sawdust collected by drilling with an electric bore ca. 1–2 cm into the trunk (bore diameter 0.5 cm). For the DNA extraction, we used a modified ATMAB protocol [43]. For the genotyping we used three chloroplast microsatellite (cpSSR) markers *ucd_4*, *udt_4* and *udt_1* [44] (Table 2). The cpSSR primers were selected based on the finding that the pair of markers *ucd4* and *udt4* is sufficient for distinguishing between the main postglacial migration lineages detectable by PCR-RFLP by [13,20].

Table 2. Description of the cpSSR loci used in our study.

Locus	Primer Sequence (5′–3′)	Repeat	Size, bp	Na	Ta (°C)	GeneBank Accession id, Genomic Location
udt1	NED-ATCTTACACTAAGCTCGGAA TTCAATAACTTGTTGATCCC	$(A)_{11}$	80–81	2	48	AJ489829, intergenic trnE-trnT
udt4	FAM-GATAATATAAAGAGTCAAAT CCGAAAGGTCCTATACCTCG	$(A)_9$	141–143	3	44 TD	AJ489831, Intergenic trnE-trnT
ucd4	FAM-TTATTTGTTTTTGGTTTCACC TTTCCCATAGAGAGTCTGTAT	$(T)_{12}$	92–95	4	45 TD	AJ489837, Intergenic ycf6-psbM

The allele number (Na) and the fragment size are given as found in our study. Ta is the primer annealing temperature (TD means touchdown used). All primers from [45].

The DNA concentration taken for the PCR was adjusted to approximately 30–50 ng/µL. The PCR amplification was carried in a 15 µL reaction mix containing 7.5 µL of the Platinum™ Multiplex PCR Master Mix (ThermoFisher Scientific, Waltham, MA, USA), 3 µL of RNAse free water, 1.5 µL of Primer Mix (2 µM each primer), 1 µL of DNA, 1 µL of PVP (1%) and 1 µL of BSA 20 mg/mL. The PCR was run with Applied Biosystems (Waltham, MA, USA) thermo cycler GeneAmp PCA System 9700 as follows: an initial pre-denaturation step at 95 °C for 15 min, followed by 26 cycles of denaturation at 95 °C for 30 s, annealing at 57 °C for 1 min 30 s, and extension at 72 °C for 30 s, followed by the final extension step at 60 °C for 30 min. Failed or unclear amplification was repeated. The fragments were separated by performing a capillary electrophoresis on the ABI PRISM™ 310 genetic analyzer. The alleles were scored by the aid of a binset in GENEMAPPER soft. v4.1 [45].

We assigned the postglacial migration lineages from [19] to our cpSSR haplotypes by synchronizing our SSR allele scores at the two loci *ucd4* and *udt4* with those from [20,23] by alignment of the most common alleles as shown in Table 3. In [23] the cpSSR haplotype association with the [19] lineages is defined by analyzing the [19] reference samples in a single plate with the studied samples (Table 3).

Table 3. Synchronizing the alleles for the two cpSSR loci from our and the two other studies on *Q. robur* with the aim to assign the cpSSR haplotypes to the postglacial migration lineages by [19]. *Q. robur.* The table includes all alleles found at the loci in all three studies.

Reference (Origin of the Material)	Alleles at Locus *ucd4*, bp				Size Shift [1]	Alleles at Locus *udt4*, bp			Size Shift [1]
Our study (Lithuania)	92	93 *	94	95	Reference.	141	142 *	143	Reference
Gailing et al. 2007, (north-western Germany)	93	94 *	95	96	+1 bp	143	144 *	145	+2 bp
Neophytou and Michiels 2013 (south-western Germany)	92	93 *	94	95	0 bp	142	143 *	144	+1 bp

[1] Size shift is accounted for the alleles in our study as the reference (+1 bp means that an allele is larger for 1 bp in comparison to the allele in our study). * most common allele at each locus in the three studies.

We constructed the multilocus haplotypes and calculated the haplotype frequency for each population. The genetic differentiation among the populations was tested by AMOVA based on genetic distances with GENALEX v.6.5 [46]. The haplotype geographical distribution was visualized by displaying the haplotype proportions for each population on the map of Lithuania. To analyze the geographical structure based on the allele frequencies, we used the Monmonier's algorithm allowing for establishing barriers along a significant shift in the allele frequency within a landscape implemented in soft. BARRIER v.2 [47]. The programme (a) creates a Delaunay triangulation plot between the sampled populations, (b) calculates genetic distances (DA genetic distance [48] in our case) associated with each edge in the plot, and (c) creates growing barriers along the largest genetic distances on the plot, the barriers are ranked based on the magnitude of the differentiation.

3. Results

All three cpSSR loci were polymorphic with two to four alleles yielding eight different haplotypes (Table 4). Based on the alignment for haplotype identity, we found the representatives of all three main postglacial migration lineages A, B and C in Lithuania (Figure 2). Most of the haplotypes (74.5%) originated from the Balkan refugium (coded as Haplo-3 in our study, corresponding to lineage A [19] or haplotypes H10-12 as in [14]). The haplotypes Haplo-3 and Haplo-4 formed the second most abundant haplogroup with 11.5% frequency. This haplogroup originates from the Balkans but outbranched from the main lineage by a mutation in central Europe (likely present-day Germany [20]). The haplotype Haplo-6 with 6.4% frequency represents the eastern Italian lineage common in Belarus and western parts of European Russia (the lineage C, H2). The western Italian haplogroup (the lineage C, H1, in our study the haplotypes Haplo-7 and Haplo-8) occurred at 4.4% frequency. Haplo-1 was found in a single individual in ROKI population but was absent in [20,23] (Table 4).

Table 4. The haplotype frequencies and descriptions. The allele sizes are given as found in our study. Frequency is the number of trees carrying a particular haplotype.

Haplotype	Frequency	*ucd_4* Allele Size, bp	*udt_4* Allele Size, bp	*udt_1* Allele Size, bp	Postglacial Migration Lineages and Haplotypes Based on [14,19,20,23]
Haplo-1	1	93	141	81	[20,23] not found
Haplo-2	4	94	141	80	H10-11-12 (lineage B) common SP, FR found in north DE and eastern SE
Haplo-3	117	93	142	81	H5 ir H7 (lineage A) north-western Balkans
Haplo-4	7	94	142	80	H4 (line A) post glacial mutation common in DE PL, HU, RO
Haplo-5	11	94	142	81	H4 (lineage A) post glacial mutation common in DE PL, HU, RO
Haplo-6	10	92	143	81	H2 (lineage C) eastern branch of Italian refugium
Haplo-7	1	94	143	80	H1 (lineage C) western branch from Italian refugium
Haplo-8	6	94	143	81	H1 (lineage C) western branch from Italian refugium

The AMOVA revealed very strong population genetic differentiation based on the cpSSR markers with the PhiPT value as high as 0.3 (the *p*-value < 0.001) and population variance component of 30% of the total variation in the molecular data (Table 5).

Table 5. The results of AMOVA on cpSSR alleles. The *p*-value of the differentiation index was based on 9999 permutations.

Source	df	SS	MS	Est. Var.	%
Among Pops	30	27.0	0.90	0.123	30%
Within Pops	127	36.1	0.29	0.292	70%
Total	157	63.2		0.414	100%

Stat	Value	*p* (rand $\geq$ data)			
Phi$_{PT}$	0.30	0.0001			

The geographical structuring of the chloroplast DNA haplotypes in the country clearly was not random (Figure 2). The most frequent Balkan lineage A (Haplo-3 in our study) was found in most of the populations with relatively higher proportions in the southern part of the country (Haplo-3 in white, Figure 2). The "German" branch of the Balkan lineage A (Haplo-4-5 in our study) spread mainly into northern Lithuania only. The eastern Italian lineage C was found exclusively in the southeastern part of the country (Haplo-6 in black, Figure 2). The western Italian branch of the C lineage, however, was exclusively present in north-eastern Lithuania (Haplo-7 and Haplo-8 in blue diagonals, Figure 2). Finally, in a single seaside population of JUOD, we found four individuals originating from as far as the Iberian lineage B (Haplo-2 in yellow, Figure 2). According to the BARRIER analysis, the following main geographically consistent gradients for significant allele frequency shifts were observed (in the order of decreasing significance): (1) separating the southeastern Lithuanian *Q. robur* populations (likely due to the presence of the eastern Italian lineage C), (2) separating western Lithuania (likely due to western Italian C and Iberian B lineages) and (3) separating north-eastern Lithuania (likely due to "German" sub-branch of Balkan lineage A) (Figure 2).

4. Discussion

We used the most frequent allele at each locus to synchronize the oak cpSSR haplotypes with those in [20,23] (Table 3). It is a subjective method, however, the following arguments indicate that we may be near to the correct associations: (a) the most frequent Balkan linage A in our study was also most frequent in [20,23], (b) the number alleles in the *ucd4* and *udt4* loci was the same as in [23], and (c) for the capillary electrophoresis and the allele scoring, we used the same hardware and software as in [20,23].

As in the previous studies on the Baltic oaks [10,22], we found all three main evolutionary lineages of oaks in Lithuania, among which the Balkan lineage A was the most frequent, and the geographical distribution of the haplotypes was in good agreement with the two above mentioned studies. This lineage diversity confirms repeated oak recolonization events into the eastern Baltic Sea region as discussed by [10]. The national gene conservation program could be adjusted to capture each lineage by at least one genetic reserve.

The geographical barriers of significant shifts in the maternally inherited *Q. robur* chloroplast DNA allele frequencies among *Q. robur* populations in Lithuania (Figure 3) correspond well to the geographical arrangement of the haplotypes (Figure 2). Basically, these barriers outlined the western Italian lineage C in western Lithuania, the "German" sub-branch of Balkan lineage A in north-east and eastern Italian lineage C in the southeast of Lithuania by leaving central Lithuania for the dominance of the Balkan lineage A (Figure 2).

The geographical distribution of the maternally inherited *Q. robur* haplotypes in Lithuania clearly has a pattern. This pattern raises questions on why in the light of large geographical scale of evolutionary migration events into such small territory (ca. 65 K km^2), there is such discrete geographical structuring of the maternally inherited chloroplast DNA haplotypes? The first question is why the Balkan lineage A is the most widespread and frequent in Lithuania? The Balkan lineage A was also the most frequent cpDNA haplotype in Lithuania based on the PCR-RLFP genotyping by [22]. It is likely that owing to the shortest geographical distance and perhaps favorability of the northward migration flow of major rivers such like Wisla, Order, and Nemunas, the Balkan refugium lineage had an advantage compared to its nearest competitors from the Italian lineage C. In addition, the geographical positioning of the Alps is less favorable for northwards migration from the Italian refugia than that of Carpathians for migration from the Balkan refugia. In agreement with [22], we found the eastern Italian Lineage C in eastern Lithuania. This result may be interpreted as a relatively later north-westwards migration of the eastern Italian lineage from the south-eastern European plains of Russia [21]. It is likely that the eastern Italian lineage C had recently reached the Baltic region, because it is rare elsewhere westwards in Lithuania (Figure 2). This finding may help to interpret postglacial migration rates.

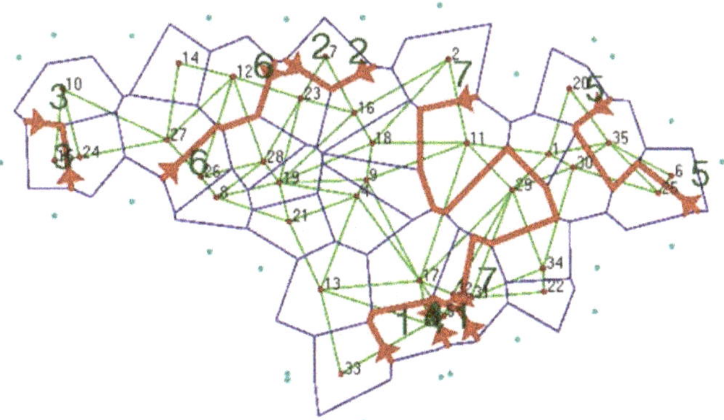

Figure 3. Results of the Monmonier's algorithm modeling (the BARRIER software) of the geographical barriers for significant shifts in the allele frequencies based on the chloroplast microsatellite markers. The map shows a Delaunay triangulation plot between the sampled populations on a simplified map of Lithuania. The larger numbers at the barriers indicate the significance of the barriers with the 1st barrier being the most significant and the 7th barrier—least significant. The small numbers at the dots indicate populations.

We found the haplotypes of the Iberian lineage B in four trees only, all located in the seaside spit of Neringa. The other studies reported the Iberian haplotypes of *Q. robur* in north-western Poland (close to the southern connection of the seaside spit to the continent) and eastern Sweden [10,19]. Thus, the samples of the lineage B we found in Neringa could be a continued natural migration from the Polish populations northwards, or less likely bird mediated travelers over the Baltic sea from Sweden. Additionally, there is a possibility of artificial introduction of oaks of the lineage B, because the seaside spit of Neringa formerly was a part of Eastern Prussia, where introduction of forest trees from other parts of the German state were common [49].

The geographical distribution of the remaining haplotypes, however, is less easy to explain by natural processes of migration alone. The first unexpected result under natural migration is a very compact concentration of the western Italian lineage C in the western highlands of Zemaitija. If considering natural migration of this haplotype northwards from Poland via south-western Lithuania, it should have entered Lithuania and spread over a much broader range from southern to western Lithuania. However, with a comparably dense sampling grid we were able to detect this western Italian lineage only in the very north-western part of the country. The second, seemingly odd geographical distribution concerns the "German" sub-branch of the Balkan lineage A. Based on the earlier studies [14,20], this sub-branch of the Balkan lineage arose somewhere in present-day Germany and we assume that it migrated from Germany north-eastwards into Lithuania. If considering a natural migration pathway, as with the western Italian lineage, we should be able to find more representatives of the "German" sub-branch of the Balkan lineage in the southern and central parts of Lithuania. However, this was not the case in our study.

What could be the main reasons for such a geographically biased distribution of the western Italian and the "German" lineages in Lithuania? An explanation could be different forestry practices and seed sources used in the former Prussian and Russian imperial states [50], as is the case of Slavonian oak introduced into North Rhine-Westphalia of Germany in the second half of the 19th century [20]. However, this is unlikely to be the main reason because the former East Prussian province occupied the very seaside and south-western parts of Lithuania, far shorter distances than both the western Italian and the

"German" lineages were located in the present study. Another more feasible explanation could be commercial exploitation of oak trees with long, straight, and clear boles that intensified in Lithuania since the 16th century and on to meet the continuing demand of quality oaks for multiple purposes [2]. In support of this hypothesis [51], reported reduced differentiation among *Q. robur* populations in the region of western Europe under stronger commercial exploitation of forests. Stronger historical forest felling happened in south-western Lithuania [2], where population differentiation is relatively lower and Balkan lineage A dominates (Figure 2). This hypothesis could be true, given the presence of associations between the stem morphotype and the chloroplast DNA haplotype of oak trees. If true, then over centuries, certain commercially fit haplotypes may have gradually been eliminated in southwestern and central Lithuania owing to better trade conditions and stronger domestication of land than in the other more forested regions of the country.

Differential adaptation of the *Q. robur* haplotypes to eco-climatic gradients of Lithuania is also possible, as discussed in [22]. However, the surface of Lithuania is flat with the altitudinal rage from ca. 10 m.a.s.l. at the seaside to ca. 250 m.a.s.l. on the highlands, the longitudinal and latitudinal gradients are small, spanning ca. 350–400 km. Such gradients apparently are too narrow to cause such a discrete adaptation-based geographical structuring of cpSSR haplotypes as observed in our study. Here again, a study on the haplotype–morphotype associations in *Q. robur* would be of value.

In conclusion, we found high diversity of *Q. robur* chloroplast DNA haplotypes for such a small territory as Lithuania containing haplogroups from the four major European lineages: Iberian, western and eastern Italy, and the Balkans. The haplotypes were geographically well structured which is difficult to explain by natural migration alone. This finding is in favor of the hypothesis of commercial elimination of certain haplotypes over centuries of exploitation of oak forests in the region. Repeated migration events may have also contributed to the present-day haplotype structure. For further testing this oak overexploitation hypothesis, a study on haplotype–morphotype associations of oaks would be of value.

Author Contributions: Conceptualization, V.B., D.D.; methodology, D.D., V.B.; laboratory analysis, J.B. and R.K.; resources, V.B.; data curation, D.D.; writing—original draft preparation, D.D. All authors have read and agreed to the published version of the manuscript.

Funding: The study was financially supported by the Ministry of Environment, the long-term Vytautas Magnus university research project "Genetic stability of forest tree species and response to environmental stress", and the long-term LAMMC research program "Sustainable Forestry and Global Change".

Data Availability Statement: The data is available upon request by e-mail to darius.danusevicius@vdu.lt.

Conflicts of Interest: The authors declare no conflict of interest.

References

1. Eaton, E.; Caudullo, G.; Oliveira, S.; de Rigo, D. Quercus robur and *Quercus petraea* in Europe: Distribution, habitat, usage and threats. In *European Atlas of Forest Tree Species*; San-Miguel-Ayanz, J., de Rigo, D., Caudullo, G., Houston Durrant, T., Mauri, A., Eds.; Publications Office of the European Union: Luxembourg, 2016.
2. Karazija, S. (Ed.) Historical Review of Oaks Forests in Lithuania. In *Oak Stands in Lithuania: Protection and Reforestation Problems*; Lithuanian Forest Research Institute: Kaunas, Lithuania, 1997; pp. 13–17. (In Lithuanian)
3. Ducousso, A.; Bordacs, S. *EUFORGEN Technical Guidelines for Genetic Conservation and Use for Pedunculate and Sessile Oaks (Quercus robur and Q. petraea)*; International Plant Genetic Resources Institute: Rome, Italy, 2004.
4. Kremer, A.; Petit, R.J. Gene diversity in natural populations of oak species. *Ann. Sci. For.* **1993**, *50*, 186–202. [CrossRef]
5. Tomiczek, C. Oak decline in Austria and Europe. *J. Arboric.* **1993**, *19*, 71–73.
6. Gustiene, A.; Vasiliauskas, A. Oak decline in Lithuania: Preliminary estimating of extend of the process and role of fungi. *Miskininkyste* **2006**, *2*, 5–11, (In Lithuanian with English summary, Figure, Table headings).
7. Finch, J.P.; Brown, N.; Beckmann, M.; Denman, S.; Draper, J. Index measures for oak decline severity using phenotypic descriptors. *For. Ecol. Manag.* **2021**, *485*, 118948. [CrossRef]
8. Reed, K.; Brown, N.; Vanguelova, E.; Finch, J.; Denman, S. Current Understanding of Acute Oak Decline: A summary of recent research with implications for woodland management. *Q. J. For.* **2021**, *115*, 38–45.

9. Aldrich, P.R.; Glaubitz, J.C.; Parker, G.R.; Rhodes, O.E.; Michler, C.H. Genetic Structure Inside a Declining Red Oak Community in Old-Growth Forest. *J. Hered.* **2005**, *96*, 627–634. [CrossRef]

10. Csaikl, U.M.; Glaz, I.; Baliuckas, V.; Petit, R.J.; Jensenen, J.S. Chloroplast DNA variation of white oak in the Baltic countries of Poland. Chloroplast DNA variation of white oak in the Baltic countries of Poland. *For. Ecol. Manag.* **2002**, *156*, 211–222. [CrossRef]

11. Dumolin, S.; Demesure, B.; Petit, R.J. Inheritance of chloroplast and mitochondrial genomes in pedunculate oak investigated with an efficient PCR method. *Theor. Appl. Genet.* **1995**, *91*, 1253–1256. [CrossRef]

12. Clegg, M.T.; Brandon, S.G.; Learn, G.H.; Morton, B. Rates and patterns of chloroplast DNA evolution. *Proc. Natl. Acad. Sci. USA* **1994**, *91*, 6795–6801. [CrossRef] [PubMed]

13. Petit, R.J.; Kremer, A.; Wagner, D.B. Geographic structure of chloroplast DNA polymorphisms in European oaks. *Theor. Appl. Genet.* **1993**, *87*, 122–128. [CrossRef] [PubMed]

14. Dumolin-Lapegue, S.; Demesure, B.; Le Corre, V.; Fineschi, S.; Petit, R.J. Phylogeographic structure of white oaks throughout the European continent. *Genetics* **1997**, *146*, 1475–1487. [CrossRef]

15. Huntley, B.; Birks, H.J.B. *An Atlas of Past and Present Pollen Maps for Europe 0–13,000 Years Ago*; Cambridge University Press: Cambridge, UK, 1983.

16. Tollefsrud, M.M.; Sønstebø, J.H.; Brochmann, C.; Johnsen, Ø.; Skrøppa, T.; Vendramin, G.G. Combined analysis of nuclear and mitochondrial markers provide new insight into the genetic structure of North European *Picea abies*. *Heredity* **2009**, *102*, 549–562. [CrossRef]

17. Pigott, C.D. *Lime-Trees and Basswoods: A Biological Monograph of the Genus Tilia*; Cambridge University Press: Cambridge, UK, 2012.

18. Danusevičius, D.; Kembrytė, R.; Buchovska, J.; Baliuckas, V.; Kavaliauskas, D. Genetic signature of the natural genepool of *Tilia cordata* Mill. in Lithuania: Compound evolutionary and anthropogenic effects. *Ecol. Evol.* **2021**, (in press). [CrossRef]

19. Petit, R.J.; Csaikl, U.M.; Bordács, S.; Burg, K.; Coart, E.; Cottrell, J.; van Dam, B.; Deans, J.D.; Dumolin-Lapègue, S.; Fineschi, S.; et al. Chloroplast DNA variation in European white oaks phylogeography and patterns of diversity based on data from over 2600 populations. *For. Ecol. Manag.* **2002**, *156*, 5–26. [CrossRef]

20. Gailing, O.; Wachter, H.; Heyder, J.; Schmitt, H.P.; Finkeldey, R. Chloroplast DNA analysis in oak stands (*Quercus robur* L.) in North Rhine-Westphalia with presumably Slavonian origin: Is there an association between geographic origin and bud phenology? *J. Appl. Bot. Food Qual.* **2007**, *81*, 165–171.

21. Semerikova, S.A.; Isakov, I.Y.; Semerikov, V.L. Chloroplast DNA Variation and Phylogeography of Pedunculate Oak *Quercus robur* L. in the Eastern Part of the Range. *Russ. J. Genet.* **2021**, *57*, 47–60. [CrossRef]

22. Pliura, A.; Rungis, D.; Baliuckas, V. Population structure of pedunculate oak in Lithuania based on analysis of chloroplast DNA haplotypes and adaptive traits. *Balt. For.* **2009**, *15*, 212.

23. Neophytou, C.; Michiels, H.G. Upper Rhine Valley: A migration crossroads of middle European oaks. *For. Ecol. Manag.* **2013**, *304*, 89–98.

24. Huntley, B. European post-glacial forests: Compositional changes in response to climatic change. *J. Veg. Sci.* **1990**, *1*, 507–518. [CrossRef]

25. Kvist, L. Phylogeny and Phylogeography of European Parids. Ph.D. Thesis, University of Oulu, Oulu, Finland, 2000. Available online: http://urn.fi/urn:isbn:9514255364 (accessed on 13 March 2021).

26. Giesecke, T.; Brewer, S.; Finsinger, W.; Leydet, M.; Bradshaw, R.H.W. Patterns and dynamics of European vegetation change over the last 15,000 years. *J. Biogeogr.* **2017**, *44*, 1441–1456. [CrossRef]

27. Buchovska, J.; Danusevicius, D.; Stanys, V.; Šikšnianienė, J.B.; Kavaliauskas, D. The location of the northern glacial refugium of Scots pine based on mitochondrial DNA markers. *Balt. For.* **2013**, *19*, 2–12.

28. Birks, H.J.B.; Tinner, W. Past forests of Europe. In *European Atlas of Forest Tree Species*; San Miguel-Ayanz, J., de Rigo, D., Caudullo, G., Houston Durrant, T., Mauri, A., Eds.; Publications Office of the European Union: Luxembourg, 2016; p. 1045.

29. Balakauskas, L. Forest Vegetation During Late Glaciation Holocene in Lithuania Mased on LRA Modeling. Ph.D. Thesis, Vilnius University, Vilnius, Lithuania, 2012. (In Lithuanian with English summary).

30. Žulkus, V.; Girininkas, A. The eastern shores of the Baltic Sea in the Early Holocene according to natural and cultural relict data. *Geo: Geogr. Environ.* **2020**, *7*, e00087. [CrossRef]

31. Kremer, A.J.; Kleinschmit, J.; Cottrell, E.P.; Cundall, J.D.; Deans, A.; Ducousso, A.O.; Konig, A.J.; Lowe, R.C.; Munro, R.J.; Petit, B.R.S. Is there a correlation between chloroplastic and nuclear divergence, or what are the roles of history and selection on genetic diversity in European oaks? *For. Ecol. Manag.* **2002**, *156*, 75–87. [CrossRef]

32. Petit, R.J.; Bodenes, C.; Ducousso, A.; Roussel, G.; Kremer, A. Hybridization as a mechanism of invasion in oaks. *New Phytol.* **2004**, *161*, 151–164. [CrossRef]

33. Bussotti, F.; Grossoni, P. European and Mediterranean oaks (*Quercus*, L.; Fagaceae): SEM Characterization of the Micromorphology of the Abaxial Leaf Surface. *Bot. J. Linnaean Soc.* **1997**, *124*, 183–199. [CrossRef]

34. Dupouey, J.L.; Badeau, V. Morphological Variability of Oaks (*Quercus robur* L., *Quercus petraea* (Matt) Liebl, *Quercus pubescens* Willd.) in Northeastern France: Preliminary Results. *Ann. For. Sci.* **1993**, *50*, 35s–40s. [CrossRef]

35. del Río, S.; Álvarez, R.; Candelas, A.; González-Sierra, S.; Herrero, L.; Penas, A. Preliminary Study on Taxonomic Review Using Histological Sections of Some Iberian Species from the Genus *Quercus* L. (Fagaceae). *Am. J. Plant Sci.* **2014**, *5*, 2773–2784.

36. Enescu, C.M. A Dichotomous Determination Key for Autochthonous Oak Species from Romania. *J. Hortic. For. Biotechnol.* **2017**, *21*, 58–62.

37. Rushton, B.S. Natural Hybridization within the Genus *Quercus* L. *Ann. For. Sci.* **1993**, *50*, 73–90. [CrossRef]
38. Schicchi, R.; Mazzola, P.; Raimondo, F.M. Eco-Morphologic and Taxonomic Studies of *Quercus* Hybrids (*Fagaceae*) in Sicily. *Bocconea* **2001**, *13*, 485–490.
39. Tamošaitis, S.; Jurkšienė, G.; Petrokas, R.; Buchovska, J.; Kavaliauskienė, I.; Danusevičius, D.; Baliuckas, V. Dissecting Taxonomic Variants within *Ulmus* spp. Complex in Natural Forests with the Aid of Microsatellite and Morphometric Markers. *Forests* **2021**, *12*, 653. [CrossRef]
40. Danusevicius, D.; Marozas, V.; Brazaitis, G.; Petrokas, R.; Christensen, K.I. Spontaneous hybridization between *Pinus mugo* (Turra) and *Pinus sylvestris* (L.) at the Lithuanian sea-side: A morphological survey. *Sci. World J.* **2012**, *2012*, 172407. [CrossRef] [PubMed]
41. Danusevičius, D.; Buchovska, J.; Stanys, V.; Šikšnianiene, J.B.; Marozas, V.; Bendokas, V. DNA marker based identification of spontaneous hybrids between *Pinus mugo* and *Pinus sylvestris* at the Lithuanian sea-side. *Nord. J. Bot.* **2013**, *31*, 001–009. [CrossRef]
42. *Lithuanian Forest Statistics Yearbook for 2019*; State Forest Service: Kaunas, Lithuania, 2020.
43. Dumolin-Lapègue, S.; Pemonge, M.H.; Gielly, L.; Taberlet, P.; Petit, R.J. Amplification of oak DNA from ancient and modern wood. *Mol. Ecol.* **1999**, *8*, 2137–2140. [CrossRef] [PubMed]
44. Deguilloux, M.F.; Dumolin-Lapègue, S.; Gielly, L.; Grivet, D.; Petit, R.J. A set of primers for the amplification of chloroplast microsatellites in *Quercus*. *Mol. Ecol. Notes* **2003**, *3*, 24–27. [CrossRef]
45. *User Bulletin GeneMapper®Software*; Version 4.0; Applied Biosystems: Foster City, CA, USA, 2006.
46. Peakall, R.; Smouse, P.E. GENALEX 6: Genetic analysis in Excel. Population genetic software for teaching and research. *Mol. Ecol. Notes* **2006**, *6*, 288–295. [CrossRef]
47. Manni, F.; Guerard, E.; Heyer, E. Geographic Patterns of (Genetic, Morphologic, Linguistic) Variation: How Barriers Can Be Detected by Using Monmonier's Algorithm. *Hum. Biol.* **2004**, *76*, 173–190. [CrossRef] [PubMed]
48. Nei, M.; Tajima, F.; Tateno, Y. Accuracy of estimated phylogenetic trees from molecular data. II. Gene frequency data. *J. Mol. Evol.* **1983**, *19*, 153–170. [CrossRef]
49. Kembrytė, R.; Danusevičius, D.; Buchovska, J.; Baliuckas, V.; Kavaliauskas, D.; Fussi, B.; Kempf, M. DNA-based tracking of historical introductions of forest trees: The case of European beech (*Fagus sylvatica* L.) in Lithuania. *Eur. J. For. Res.* **2021**, *140*, 435–449. [CrossRef]
50. Deltuvas, R. *Forestry and hunting in Little Lithuania. Monography*; Klaipeda University Press: Klaipėda, Lithuania, 2019.
51. König, A.O.; Ziegenhagen, B.; van Dam, B.C.; Csaikl, U.M.; Coartd, E.; Degena, B.; Burg, K.; de Vries, S.M.G.; Petita, R.J. Chloroplast DNA variation of oaks in western Central Europe and genetic consequences of human influences. *For. Ecol. Manag.* **2002**, *156*, 147166. [CrossRef]

forests

MDPI

Article

An Updated Infrageneric Classification of the North American Oaks (*Quercus* Subgenus *Quercus*): Review of the Contribution of Phylogenomic Data to Biogeography and Species Diversity

Paul S. Manos [1,*] and **Andrew L. Hipp** [2]

[1] Department of Biology, Duke University, 330 Bio Sci Bldg, Durham, NC 27708, USA
[2] The Morton Arboretum, Center for Tree Science, 4100 Illinois 53, Lisle, IL 60532, USA; ahipp@mortonarb.org
[*] Correspondence: pmanos@duke.edu

Abstract: The oak flora of North America north of Mexico is both phylogenetically diverse and species-rich, including 92 species placed in five sections of subgenus *Quercus*, the oak clade centered on the Americas. Despite phylogenetic and taxonomic progress on the genus over the past 45 years, classification of species at the subsectional level remains unchanged since the early treatments by WL Trelease, AA Camus, and CH Muller. In recent work, we used a RAD-seq based phylogeny including 250 species sampled from throughout the Americas and Eurasia to reconstruct the timing and biogeography of the North American oak radiation. This work demonstrates that the North American oak flora comprises mostly regional species radiations with limited phylogenetic affinities to Mexican clades, and two sister group connections to Eurasia. Using this framework, we describe the regional patterns of oak diversity within North America and formally classify 62 species into nine major North American subsections within sections *Lobatae* (the red oaks) and *Quercus* (the white oaks), the two largest sections of subgenus *Quercus*. We also distill emerging evolutionary and biogeographic patterns based on the impact of phylogenomic data on the systematics of multiple species complexes and instances of hybridization.

Keywords: biodiversity; distribution; hybridization; oak; phylogeny; systematics

Citation: Manos, P.S.; Hipp, A.L. An Updated Infrageneric Classification of the North American Oaks (*Quercus* Subgenus *Quercus*): Review of the Contribution of Phylogenomic Data to Biogeography and Species Diversity. *Forests* **2021**, *12*, 786. https://doi.org/10.3390/f12060786

Academic Editors: Mary Ashley and Janet R. Backs

Received: 5 May 2021
Accepted: 10 June 2021
Published: 15 June 2021

Publisher's Note: MDPI stays neutral with regard to jurisdictional claims in published maps and institutional affiliations.

1. Introduction

The oaks (*Quercus*, Fagaceae) comprise more than 435 species distributed across temperate and tropical regions of the northern hemisphere [1–3]. They are among the best-known and most ecologically significant forest trees in North America (including Mexico) and Eurasia, shaping forest and savanna ecosystems [4] and the diversity of urban forests [5,6], and molding the development of human civilization and mythology [7–10]. Oaks are united by a suite of distinctive floral traits—in particular, pendant male catkins and tricarpellate female flowers with linear styles and expanded stigmas—and a signature fruit trait: the circular single fruit or acorn seated within a cup-shaped extra-floral accessory structure, a specialized involucre called a cupule. The cupule is the defining feature of Fagaceae. While other genera of the family feature acorns and valveless rounded cupules, the combination of floral character states shared by oaks is unique within the family [11,12].

The most recent classification system of *Quercus* recognizes eight sections within two subgenera, subgenus *Quercus* (ca. 295 spp.) and subg. *Cerris* (ca. 140 spp. [3]; see Table 1). This old dichotomy between oak subgenera, which dates to the early Eocene, was recognized in early molecular studies of the genus and family [13,14] and is robustly supported by an extensive RAD-seq (next-generation DNA sequencing) phylogenomic study of a comprehensive worldwide sample of species [15]. The two subgenera represent a deep biogeographic split among modern taxa: subg. *Cerris* is restricted to Europe, Asia, and northern Africa; and subg. *Quercus*, the American oak clade, is largely restricted to the Americas except for two dispersals back to Eurasia. Eight sections of the genus are also supported by a combination of RAD-seq data and macro- and microscopic characters, most notably patterns of pollen ornamentation viewed under SEM [3]. These subtle yet well-established differences in pollen exine, and their correlation with phylogenomic data, give us high confidence in the basic split between subgenera and the pollen subtypes that correspond to each of the recognized sections [3,16].

Table 1. Current Infrageneric classification of the genus *Quercus* [3]. The total number of species with in each infrageneric group is provided, including the number of species within the Flora of North America (FNA) region and the number treated within this study.

Quercus L.	Sections	No. of Species Total/FNA	No. of Species Treated Here
subg. *Quercus*		297/92	
	Lobatae Loud.	120/39	37
	Protobalanus (Trel.) Schwarz	5/5	
	Ponticae Stef.	2/1	
	Virentes Loud.	7/4	
	Quercus	150/53	45
subg. *Cerris* Oerst.		138	
	Cyclobalanopsis (Oerst.) Benth. & Hook. f.	90	
	Cerris Dumort.	13	
	Ilex Loud.	35	

Fossilized pollen has been used to pin the modern sections in time and space, serving as an invaluable resource to interpret paleobiogeography and to time-calibrate the global oak phylogeny [15,16]. Time-calibrated phylogenomic analyses, combined with fossil and modern oak species distributions, support an initial split in the American oak clade between the red oaks, sect. *Lobatae*, and the remainder of the clade between ca. 54 and 48 Ma (= million years ago). The non-*Lobatae* American oak clade sections exhibit high ecological diversity, with several striking contrasts between three early-diverging clades—sects. *Protobalanus*, *Ponticae* and *Virentes*—and the main radiation of white oaks, sect. *Quercus*. These early-branching lineages are species-poor, two to seven species each, restricted in distribution, and evergreen to wintergreen in habit. By contrast, sect. *Quercus*, with approximately 150 species of white oak, has the broadest distribution of all forest trees in the Northern Hemisphere [3]. This may be owing to the evolution of deciduousness and other traits associated with a range of continental climates, an exceptionally high range of functional diversity that sets the American oaks up as a model clade for understanding tree diversification and ecology [17,18]. The other main branch of the American oaks and the sister group to these lineages, sect. *Lobatae*, comprises the second largest clade next to the white oaks. It forms a remarkably parallel radiation to the white oaks with respect to distribution, habit, and morphology [17,19]. There is, however, an important point of ecological distinction between these two large oak clades: *Lobatae* show lower levels of diversification in western North America, especially in the California Floristic Province (CA-FP) and xeric woodlands of the American southwest [20]. The reasons for this ecological distinction between the clades may be key to understanding coexistence of these sections in the Americas, and the consequent interaction between diversification and tree diversity in oaks [17,21,22]

The sectional classification for *Quercus* [3] is solid and unlikely to change. In this paper, we present a subsectional classification for the American oak clade based largely on our phylogenomic work of the past five years [15,19,23–25]. Our current sampling of species for RAD-seq analysis stands at roughly 250 taxa: A total of 177 from subg. *Quercus*, the American oak clade, and 73 from subg. *Cerris*, the Eurasian oak clade. Within subg. *Quercus*, our coverage is nearly complete for the 92 species treated within the *Flora of North America* (FNA north of Mexico [1]). This phylogenomic work reveals compelling continental patterns of oak diversity at the broadest and narrowest phylogenetic scales. At the scale of entire clades, the early-branching clades as well as many other lineages within the two largest sections of subg. *Quercus*, sects. *Lobatae* (red oaks) and *Quercus* (white oaks), are distributed exclusively within North America north of Mexico (hereafter, "the FNA region"). It is also clear that the number of species within the FNA region with relationships to species outside the region is relatively low, limited to a single species of sect. *Ponticae* and a single clade in sect. *Quercus*, as discussed below. Thus, the oak flora of the FNA region is nearly a self-contained unit for study. At the same time, within regional clades, multiple radiations of taxa have been sampled at the population level for both our global phylogeny and as part of earlier stand-alone studies targeting the fine-scale systematics of challenging species complexes [23,25–28]. These enable the same dataset that informs classification to provide insight into species taxonomy. With this framework in place, we are well positioned to provide a phylogenetically-driven and timely bookend on the systematics of the taxa treated in the FNA region.

Our goals with this review are: (1) To describe the biogeographic patterns of oak diversity within North America; (2) to formally classify 62 species into nine well-sampled monophyletic subsections; and (3) to review the impact of phylogenomic data on the systematics of multiple species complexes and distill the results of hypothesis testing involving hybridization within the flora.

1.1. Phylogeny and Biogeography of the Oak Flora in North America

The American oak clade (*Quercus* subg. *Quercus*) comprises three endemic lineages—sects. *Lobatae*, *Protobalanus* and *Virentes*—and two Holarctic or transcontinental lineages—sects. *Ponticae* and *Quercus* (Figure 1). Sections *Lobatae*, *Protobalanus*, and *Quercus* likely arose in what is now the boreal zone. Middle Eocene pollen records from Axel Heiberg Island (79°55′N, 88°58′ W) indicate the presence of modern oak lineages at high latitudes in North America by c. 45 Ma [29]. As temperatures at high latitudes decreased by ca. 3–5 °C during the Eocene–Oligocene climate transition 34 Ma [30], *Lobatae* and *Quercus* were pushed southward toward their modern-day distributions (Figure 1). A parallel pattern of vicariance in the red oaks and the white oaks is evident from the phylogeny, with each section diverging into a western clade centered in the California Floristic Province (CA-FP) sister to an eastern North American clade. The red oaks and white oaks then radiated simultaneously in the CA-FP and eastern North America before diverging from an eastern North American ancestor to diversify in Mexico and Central America.

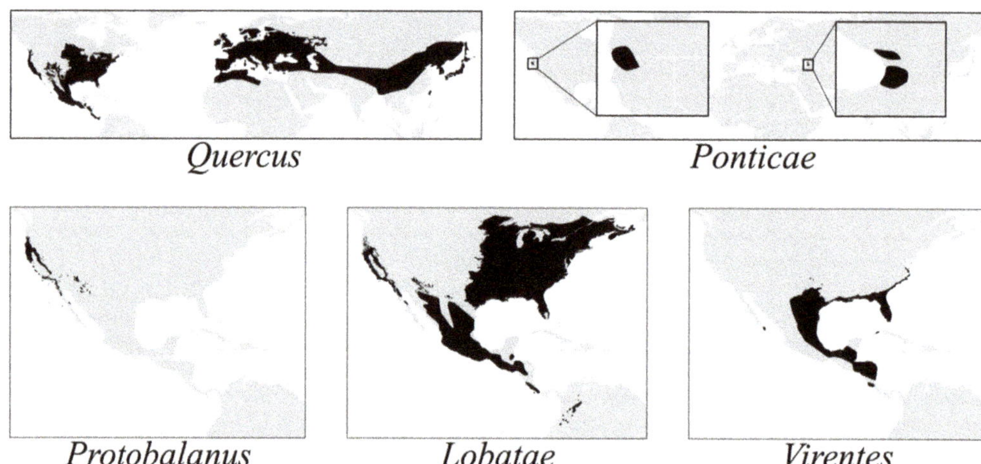

Figure 1. Distributions of the five oak sections of the American oak clade (*Quercus* subg. *Quercus*). Maps were generated using species maps provided by Little [31,32], as refined by recent studies [17] and herbarium records.

The oak flora of the FNA region contains a high level of global phylogenetic diversity—all five sections of subg. *Quercus*, including the relictual sects. *Ponticae* (*Q. sadleriana* only) and *Protobalanus* (five species), which are restricted to the CA-FP or, in the case of *Q. chrysolepis* and *Q. palmeri* of sect. *Protobalanus*, range from the CA-FP to the adjacent southwest. Our work suggests that older crown group ages [15] and plastome divergences [13,33] are associated with many of these deeply divergent subclades, a pattern consistently in contrast to the more recently derived subclades containing mostly Mexican and Central American species. These nested clades contain three unrelated red oak species and two distinct groups of white oak species in the FNA region, which suggests multiple recent dispersal events or range extensions from Mexico back into the southwestern edge of the FNA region [19].

1.2. Patterns of Ecological Diversity

A broad overview of North American species diversity and community assembly published in 2018, based on tree species within U.S. Forest Service Forest Inventory and Analysis (FIA) plots [17,18], shows that FNA oak diversity peaks in the southeastern piedmont and coastal plain regions [18]. While these analyses did not include ca. 30 oak shrubs and small tree species, they demonstrate the broad distribution of oak diversity in the FNA region. What they miss most significantly is the impressive level of shrub and small tree diversity in the CA-FP and adjacent desert biomes of the southwest that accounts for about 18 species across four sections, or roughly two-thirds of the species of smaller stature within the flora. While the addition of these species changes little to the initial observation that the uplands of eastern and central North America are the least diverse areas for oaks, this diversity is important to understanding the role of oak phylogenetic diversity in a range of ecological and evolutionary contexts, including community assembly and ecosystem functions of non-forested oak-dominated ecosystems.

The broad geographic and ecological distribution of oaks across at least eight distinct North American biomes has generated a complex array of foliar adaptations, especially within the two main species-rich sections, *Lobatae* and *Quercus*. While each of the sections covers a broad ecological and geographic range, these ranges begin to subdivide and segregate regionally. Eastern North American subclades *Stellatae* (sect. *Quercus*) and *Phellos* (sect. *Lobatae*), here treated as subsections, centered in the piedmont and coastal plain, highlight previously unsuspected close relationships between lobe-leaved and non-lobe-leaved species. In the southeast, the added presence of generally non-lobe-leaved species of sect. *Virentes* (live oaks) with already diverse communities of sympatric oak species revealed a clear pattern of phylogenetic sorting and convergent evolution among species from the three sections along moisture and fire gradients [26,34].

For species placed within western clades of red and white oaks and centered on the CA-FP—subclades *Dumosae* (sect. *Quercus*) and *Agrifoliae* (sect. *Lobatae*), here treated as subsections—a combination of sclerophylly and deciduousness has presumably evolved in response to drought. In each case, phylogenies resolved the lobe-leaved deciduous species sister to more diverse clades of mostly sclerophyllous shrub species [23,27,35]. This parallel distinction between leaf habit and species diversity in the white and red oaks of the CA-FP suggests an increase in diversification rate in response to decreasing summer precipitation and general support for the hypothesis of convergent patterns of foliar evolution. Additional complexity in the evolution of character states of leaf persistence is observed within the most diverse clades of red and white oaks along abiotic gradients southward into Mexico and Central America, demonstrating sympatric and strongly parallel diversification in climatic niche and leaf habit increasing in rate with the move into Mexico [19].

1.3. Classification of the American Oak Clade: Species Level Systematics

The ecological and biogeographic segregation of species within sections suggests a biological rationale for naming subsections, beyond the practical ramifications of any classification system for the writing of keys and identification of species. At the species level, phylogenetic relationships among oaks species remain difficult to resolve with confidence, especially for a group well known for hybridization [24,36,37]. But recent studies using a range of DNA markers suggest that hybridization does not pose an insurmountable barrier to achieving that goal [24,26,36,38–41]. Population-level sampling is critical to establishing that oak species show genetic cohesion, and the use of reduced-representation next-generation sequencing data on exemplars across species ranges has provided robust evidence for the existence of species boundaries even among closely related species occurring in sympatry (e.g., [42–46]. A species level analysis of sect. *Virentes* stands out as the first oak study with complete phylogenomic sampling of any oak clade, using RAD-seq data and including multiple accessions throughout the range of each species, to generate a phylogeny and provide the context needed to guide a range of comparative investigations [26]. Since then, two more clades, treated here as subsections *Dumosae* and *Agrifoliae*, have been comprehensively sampled and analyzed [23,27,35,47].

Throughout this paper, our work rests on a concept of oak species as populations of individuals that cohere genomically in a subset of the genome that is likely shared across all members of the species, barring early-generation hybrids [48,49]. While the mechanisms maintaining species boundaries in oaks very likely differ across the range of even a single species, and while those mechanisms maintained across a species' range may vary in strength across the range, it remains the case that we are able to recognize most oak species ecologically and morphologically [50–52]. In addition, as individual cases of species pairs and species clusters are investigated with increasing amounts of molecular data, sampled from across the genome, they converge with very few exceptions on genomic clusters that accord with recognizable species [23,38,53–56]. Exceptions will no doubt result from incomplete divergence, as time since speciation is likely a key factor in resolving oak species, even with the application of multiple criteria [48]. Our work thus far has shown that clades from the FNA region often vary in their depth of relative species divergence, with several examples that suggest we are capturing various stages of diversification, ranging from isolation by distance on the California Channel Islands (e.g., sect. *Protobalanus*, subsect. *Dumosae*) to differentiation in ecological niche space in both coastal and continental regions (e.g., sect. *Virentes*, subsect. *Prinoideae*).

Analogous to the argument that has been made for plant species as a whole [56], the correlation between morphological patterns of similarity and genomic evidence for reproductive barriers suggests that our use of the species category in oaks is both meaningful and based on an integrative approach using multiple criteria [48]. Where studies have found difficulty distinguishing morphological and ecological species using molecular data (e.g., [57,58]), returning with higher numbers of markers generally recovers clusters that cleanly separate the species (e.g., [43,59]). This suggests that there may be genomic heterogeneity in *where* and *how much* of the genome is shared within species vs. among species, even if the regions of the genome that distinguish recently diverged oak species may be few and far between [60] and interspersed with regions that exchange relatively freely between species. At the same time, it gives us good reason to expect that species we are able to recognize ecologically and morphologically will also cohere genomically, so long as we look closely enough.

Moreover, in spite of taxonomic difficulties at the species level, our RAD-seq work already cited and ongoing work based on sequence-capture data [37,61] strongly support the clades we recognize as subsections in the current work (Figure 2). Given our sampling to date and the strong phylogenetic support, the timing is reasonable to recognize subsections for the FNA region. To complete the American oak clade (*Quercus* subg. *Quercus*), better sampling of the Mexican and Central American taxa is needed and ongoing.

2. Materials and Methods

Taxonomic Treatment

In the treatments that follow, all references to phylogenetic resolution are based on RAD-seq data cited here [15,19,23–27,36,62] unless otherwise indicated. Three of the American sections are relatively small and are not divided into subsections: These are sect. *Ponticae* (2 species, one of which is in the FNA region), sect. *Protobalanus* (5 species), and sect. *Virentes* (7 species, 4 of which are in the FNA region).

Numbers in parentheses indicate the number of species in the FNA region out of the total number of species estimated for the clade, following previous work [1,3]. Species indicated with an asterisk are subshrubs, shrubs, or small trees; all others are trees. Section descriptions are not here; for these, we refer the reader to the most recent classification of the genus [3]. Our subsectional descriptions are based on previous classifications, species-level treatments, and morphological studies cited below, including new observations made from herbarium specimens.

3. Results and Discussion

3.1. Section Protobalanus (5/5)

Includes: * *Q. cedrosensis* C.H. Muller, *Q. chrysolepis* Liebm., * *Q. palmeri* (Engelm.) Engelm., *Q. tomentella* Engelm., * *Q. vacciniifolia* Kellogg

The *Protobalanus* oaks are the only truly evergreen American group, with leaves persisting up to three years. Three of the species are shrubs or small trees. All species of the section are in the FNA region, and two of them extend south into Mexico. Defining features of the section include apiculate stamens and a mostly lateral position of abortive ovules [3,63]. Trichome variation is especially informative in this species complex [64]. The most widespread species, *Q. chrysolepis*, is also one of the most variable North American oaks, with juvenile leaf morphology and branching habit, sucker shoots, shade forms, or shoots of individuals in ecologically extreme habitats often resembling other species of the complex [1]. The section includes one of the two island endemic species in the FNA region, *Q. tomentella* (California Channel Islands; Isla Guadelupe, Mexico). Clonal reproduction is common in *Q. tomentella* [65] and reported to be extreme in *Q. palmeri* from the Jarupa Hills of southern California, the site of a 13,000 year-old clone [66].

Quercus cedrosensis appears to be the earliest branch in the clade, followed by *Q. palmeri*. The remaining taxa are not cleanly resolved as subclades with our current sample, suggesting a recent divergence history and likely continued gene flow between the two species pairs, *Q. chrysolepis–Q. vacciniifolia* and *Q. chrysolepis–Q. tomentella* [65,67]. Phylogeographic studies of *Protobalanus* using plastid data showed a north-south discontinuity in California, perhaps related to an ancient disjunction within the complex that limited seed dispersal [13]. More recently, Ortego et al. [67] detected two lineages within a paraphyletic *Q. chrysolepis* using RAD-seq data and population genetic analyses. Their work demonstrates the existence of a southern lineage sharing a most recent common ancestor with *Q. tomentella*, divergence between the southern and northern lineages at about 5 Ma, consistent with our fossil-calibrated estimate [15].

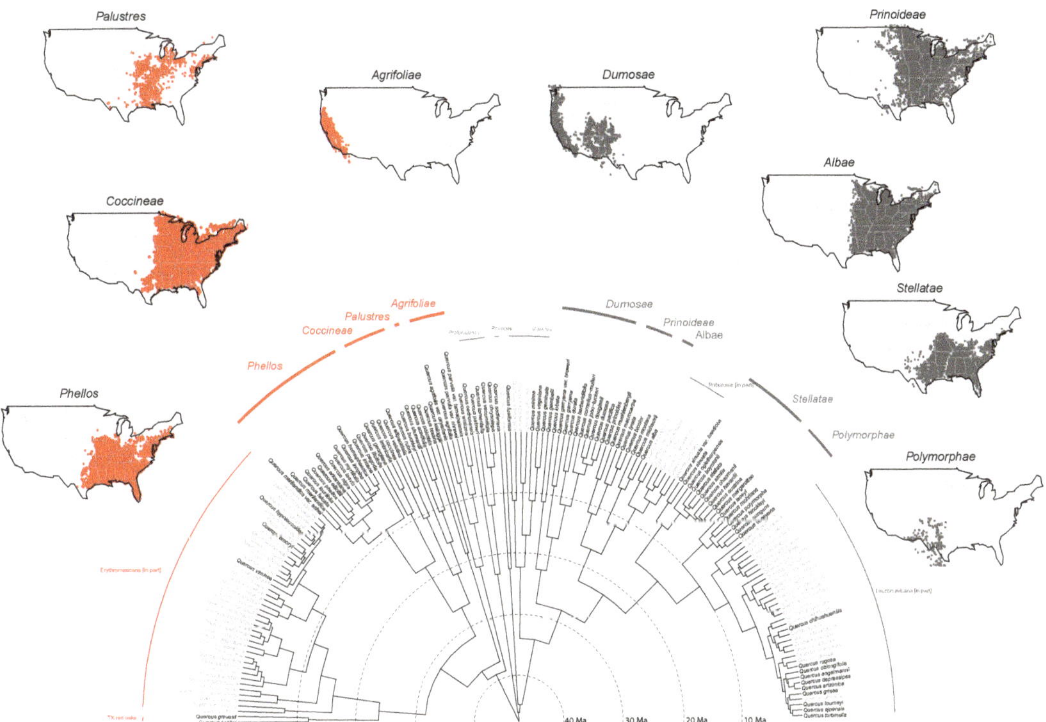

Figure 2. Time tree/chronogram of the five sections of the American oak clade (*Quercus* subg. *Quercus*) adapted from [19], with distribution maps for the nine subsections recognized within sections *Lobatae* and *Quercus*. Shown in boldface are 88 of the 92 taxa that occur within the FNA region. Ma = million years ago. Maps are based on specimen data used in [19], excluding specimen records from cultivation and from outside the range of the species; additionally, populations of *Q. havardii* from west of eastern Arizona were excluded as their taxonomic status and possible hybrid origin are uncertain [68,69]. Additional specimen data from SEINet were included for the *Coccineae* in North Carolina, which were undersampled. Code for generating maps and phylogenetic tree is archived in GitHub (https://github.com/andrew-hipp/2021-oak-classification, Release v0.95-1, 2021-06-06) and Zenodo (DOI: 10.5281/zenodo.4905230).

Both the fossil record and expanded ancestral range predicted by niche modeling for *Q. chrysolepis* suggest *Protobalanus* once were more widely distributed [70,71]. Plastome sharing between the California white oak *Q. engelmanii* and sect. *Protobalanus* and deep plastome splits between sects. *Quercus* and *Protobalanus* further support a long history of co-occurrence of the two sections in the CA-FP [13,33]. A center of diversification in western North America suggests a long-term presence in the region and perhaps farther south given the deep phylogenetic divergence of *Q. cedrosensis* and its extended range south along the Baja coast to Cedros Island, Mexico.

3.2. Section Ponticae (1/2)

Includes: * *Q. sadleriana* R. Br. Ter

Quercus sadleriana is one of the most distinctive oaks from western North America (CA-FP) and the FNA region's solitary member of this ditypic section. The combination of its isolated range and chestnut-like, shallowly lobed leaves reminiscent of the white oak species of both eastern North America and Asia likely led to its placement in monotypic groups by both Trelease ([72]; as series *Sadlerianae*) and Camus ([73–75]; subsect. *Sadlerianae*). Our analyses conclusively place it sister to *Q. pontica* K. Koch from the Caucasus Mountains, supporting a placement that had previously been shown correctly based on nuclear ribosomal DNA alone, but not robustly supported [76]. Together they form a distinct Holarctic lineage within subg. *Quercus*. *Ponticae* are quite similar in their shallowly lobed, chestnut-like leaves, large stipules, large leaf buds with loosely attached scales, and shrubby, rhizomatous habit. They also share the derived features of basal abortive ovules, glabrous endocarp, and annual fruit maturation with sects. *Virentes* and *Quercus* [3].

Qualitatively, both species occupy similar latitudes and ecological zones that overlap strongly in the mean values of 19 climatic variables [77], despite being longitudinally separated by 165°. There is morphological evidence that *Q. sadleriana* hybridizes with *Q. garryana* ([78]; A. Keuter, pers. comm.) and strong molecular evidence that ancient introgression occurred between *Q. pontica* and the roburoid white oaks of Eurasia [26,39]. This history of introgression has interfered with the accuracy of phylogenetic estimation within the white oaks and become a model for examining the impact of introgression on the mosaic of phylogenetic signals within the oak genome [15,24,37,79].

Given the nested phylogenetic position of *Ponticae* within the American oak clade, it seems reasonable to hypothesize an origin in the Americas followed by dispersal to Eurasia. Undoubtedly, *Ponticae* are relics of a once more widespread species group although there is no macrofossil evidence that can be definitively attributed to the group [3].

3.3. Section Virentes (4/7)

Includes: *Q. fusiformis* Small, *Q. geminata* Small, * *Q. minima* (Sarg.) Small, *Q. virginiana* Mill.

Although sect. *Virentes* has long been considered one of the best-demarcated species groups of the American oaks [34,80], until recently its relationship to the white oaks was unclear. All phylogenomic data place *Virentes* sister to sect. *Quercus* with strong support. Section *Virentes* is uniquely characterized by the combination of fused stellate trichomes on the leaf undersides; seeds with fused cotyledons; and the presence of a cotyledonary tube in seedling development [3,80]. Section *Virentes'* consistent and distinct morphology across vegetative and reproductive features support a long and divergent history within the American oak clade.

The lineage is unusual within the FNA region and globally in being restricted to low-elevation habitats, occurring largely on well-drained sandy soils or volcanic tuff [34]. Ecological separation of the taxa by climate and environmental factors is well studied [24], including clear demonstration of flowering time as an isolating mechanism between *Q. virginiana* and *Q. geminata* [81].

Population-level studies support the placement of the four FNA taxa into two main clades [24]. One comprises three species from the southeastern FNA region: The widespread *Q. virginiana* and more narrowly distributed sister species-pair of *Q. geminata* and *Q. minima*. These species form the sister to a Central American clade of *Q. sagraena* Nutt. and *Q. oleoides* Schltdl. and Cham. In the southwest, *Q. fusiformis* and *Q. brandegeei*, a narrow endemic from the southern tip of Baja California, diverge early from the rest of the complex. A discontinuous distribution across the southwest of the FNA region and extension south suggests allopatric speciation, driven by range contraction and isolation by distance [82]. Long distance dispersal also seems likely in explaining the origin of the Antillean *Q. sagraena* (Cuban oak) from a mainland progenitor similar to *Q. oleoides* [26].

The distribution of *Virentes* suggests physiological adaptations to temperate climates with mild winters or seasonally dry tropical climates. Vulnerability to freezing and tolerance of drought have likely influenced historical patterns of migration and the current restricted distribution of the group [17]. Although a temperate origin within the FNA region, followed by east-west dispersion and subsequent migration south seems plausible, the timing remains elusive. With no reliable fossil record for modern *Virentes* [3], the age of the crown group is dependent on sampling, phylogenetic estimation, and the use of fossil constraints derived from other oak lineages. The wide range of crown group estimates from 11 to 23 Ma is likely due to these variables [16,26], a reminder of the uncertainty and potential disparities in the timing of the pattern of oak diversification across studies.

Evidence for historical introgression between parapatric species pairs suggests hybridization during periods of range overlap or long-distance transmission of pollen [36]. For the FNA taxa, introgression among sympatric *Virentes* is clear from morphology and genomic data, but the degree of gene flow is modulated by the strength of ecological isolating mechanisms [26]. Differences in flowering time (*Q. virginiana* vs *Q. geminata*) and drought tolerance (*Q. fusiformis* vs *Q. virginiana*) work to maintain reasonably distinct genetic patterns of species cohesion. However, niche differentiation in fire response and habit between *Q. minima* and *Q. geminata*, two species with broadly overlapping flowering times, are less effective ecological barriers to gene flow. Mixed populations of *Q. virginiana*, *Q. minima*, and *Q. geminata* at the northern extent of sympatry in North Carolina appear to show elevated levels of morphological intergradation, possibly a result of habitat destruction and breakdown of ecological isolating mechanisms (P.S. Manos, pers. obs.).

4. Section Lobatae–(37/ca. 120)

Lobatae occupy a range of habitats throughout the Americas, with just one species, *Q. humboldtii* Bonpl., extending into northern Colombia [83,84]; this is the only oak species of any section to range into South America, where it is ecologically important in a diverse array of forest types [85,86]. *Lobatae* are defined by long styles encircled at their base by a perigon, a skirt-like perianth structure [3,87]. Most *Lobatae* species have biennial fruit, but three unrelated species within the FNA region bear annual fruit (*Q. agrifolia*, *Q. emoryi, and Q. elliottii*), and a fourth unrelated species, *Q. hypoleucoides*, is polymorphic [88].

The earliest branch within *Lobatae* is a clade we treat here as subsect. *Agrifoliae*, distributed largely within the CA-FP. It is sister to a succession of subclades distributed from eastern North America to well south of the FNA region. Four of these clades are densely sampled in our phylogenetic work and treated below as subsects., but a fifth group within the FNA region, informally known as the Texas red oaks (e.g., *Q. canbyi* Cory and Parks = *Q. graciliformis* C.H. Muller; = *Q. gravesii* Sudw.), may or may not contain other rare, unsampled species that occur in Texas (e.g., *Q. rubusta* C.H. Muller and *Q. tardifolia* C.H. Muller). The placement of this group between the grade formed by temperate subsects. and the large clade formed by all other *Lobatae* south of the FNA region seems to geographically parallel the white oak biogeographic patterns (Figure 2). However, additional studies are needed to evaluate the possibility that the taxonomic complexities in the Trans-Pecos Region represent relictual hybrid populations or rare taxa with relationships to species within subsect. *Coccineae* or to Mexican taxa [1,72,89,90]. With our limited sample, the Texas Red oak group is the provisional sister to the clade formed by all other *Lobatae* south of the FNA region, informally called *Erythromexicana*, whose distributions suggest at least two major hotspots of species richness: The Northern Sierra Madre Orientale and Serranías Meridionales of Jalisco [91]. Three additional species within the FNA region, *Q. emoryi* Torr., *Q. hypoleucoides* A. Camus, and *Q. viminea* Trel., are clearly related to three different subclades within *Erythromexicana*.

Reconstruction of the timing and biogeography of the *Lobatae* phylogeny suggests an initial diversification of red oaks within the FNA region, followed by movement south into Mexico about 15 million years ago, associated with an increased net diversification rate most likely attributable to increased speciation rate [19]. This increase in species diversification rate was likely a response to the combination of tectonic movement, volcanism, and climatic oscillations that generated a mosaic of seasonal montane habitats in Mexico [92,93]. One major barrier to the *Lobatae* diversification was the Nicaraguan Depression [93]. As with the parallel and equally rapid diversification of white oaks, species diversity is considerably lower southward to Costa Rica.

In this subsectional treatment, we re-classify a total of 29 species of sect. *Lobatae* recovered in our global analysis, and resolved with high confidence into four clades.

4.1. Quercus subsect. Agrifoliae *(Trel.) A. Camus, Monogr. Genre* Quercus *3: 46. 1952*

Q. subsect. *Californicae* A. Camus, Monogr. Genre *Quercus* 3: 426. 1952.

Shrubs and trees to 30 m; **Bark** gray, dark brown to black, smooth or with wrinkled rings perpendicular to trunk axis or shallowly sinuous-furrowed to deeply furrowed, occasionally cross-checked into islands; **Twigs** slender, pubescent to glabrous; **Buds** light brown to reddish brown, ovoid to conic 3–9 mm; **Leaves** deciduous to evergreen, petiole 2–40 mm; **Leaf blade** circular to elliptic, ovate, obovate to oblong or narrowly lanceolate; base obtuse, rounded to cordate; margin entire or spinose with up to 30 awns or with 7–11 acute lobes and 13–30 awns; apex acute, attenuate, blunt to rounded; surfaces abaxially glabrous to densely pubescent with multiradiate trichomes or with small axillary tufts of 4–8 rayed fasciculate trichomes; **Acorns** biennial or annual, cup thin to thick, sometimes proximally tuberculate, deeply and narrowly cup-shaped, turbinate or U-shaped to deeply bowl-shaped, scales sometimes obtuse proximally to more acute distally, tips ragged or loose especially along distal cup edge, nut ovoid, oblong, conic to broadly ellipsoid; cotyledons distinct.

Distribution: California, north to OR, south to Mexico (B.C.) (see Figure 2)

Includes: *Q. agrifolia* Née (the type), *Q. kelloggii* Newb., * *Q. parvula* Greene, * *Q. wislizeni* A.DC.

The distinct nature of the red oaks of the CA-FP was recognized by Trelease, who viewed them as an isolated lineage and treated them as series *Agrifoliae* (pages 205, 206 in [70]). His treatment was mostly followed by Camus, although she classified the deciduous species *Q. kelloggii* into a monotypic subsection. Despite the contrast in leaf habit, species of *Agrifoliae* share a suite of diagnostic features including a deep cup and oblong nut and with cup scale tips generally appearing ragged or loose and occasionally inrolled.

Agrifoliae have been the subject of recent morphometric and genetic studies, resulting in intense scrutiny of the eleven named taxa, which include four species, four varieties, and three named hybrids [22,47,94]. These studies all demonstrate that the lobe-leaved deciduous species *Q. kelloggii* is sister to the remaining subevergreen species, in support of Camus' recognition that it is taxonomically distinct. Tests for introgression based on the molecular genetic data also demonstrate that *Q. kelloggii* hybridizes with every species in the complex, but there is yet no evidence that F_1 hybrids backcross to parental species in nature [23,95]. This potential barrier to introgression, with F_1 hybrids an apparent dead-end between *Q. kelloggii* and the remainder of the section, is of broad interest: Hybridization between oak species with distinctly different leaf habits is striking on its own [96,97], and biologically, the limits on introgression within any oak section is poorly understood. We also note that the divergence time estimate between *Q. kelloggii* and the remainder of the subsection is roughly 20 Ma, among the oldest splitting times inferred within crown groups among the major oak groups within subg. *Quercus*. The deep divergence separating *Q. kelloggii* from the other California red oaks may contribute to the maintenance of reproductive isolation in the subsection. Nonetheless, at finer phylogenetic scales, lingering taxonomic problems at the varietal level in *Agrifoliae* appear to be mainly due to hybridization (see references above).

Within the subevergreen complex, *Q. agrifolia* is sister to *Q. parvula* + *Q. wislizeni*. Within each of these species, recognized varieties correspond to north-south extremes in species distribution and values in sets of continuous morphological traits [94]. For the coastally distributed *Q. parvula*, there is strong support for var. *shrevei* (C.H. Muller) Nixon as the only valid variety whereas both var. *tamalpaisensis* and var. *parvula* have complex genetic backgrounds that likely involve gene flow from different combinations of subevergreen species [47]. While more study is needed, our findings indicate that the narrowly endemic *Q. parvula* var. *tamalpaisensis* S.K. Langer represents a hybrid population that is roughly 60% *Q. wislizeni* and 40% *Q. parvula* var. *shrevei*. In this case, leaf size is notably larger in the introgressed populations, well outside the range of the parental species [23].

Disentangling *Q. parvula* var. *shrevei* from *Q. wislizeni* demonstrates that two main taxa span coastal and interior areas of the CA-FP. Molecular dating analysis [23] is consistent with the hypothesis proposed by Axelrod in 1983 [98] that *Q. parvula* and *Q. wislizeni* split from a common ancestor approximately 10–12 Ma. The fossil record shows that the *Q. wislizeni* + *Q. parvula* clade likely originated within what is now the Great Basin, east of the Sierra Nevada [99,100], with the lineage that would become modern *Q. parvula* moving westward in response to the drying climate following uplift of the Sierra Nevada. *Quercus wislizeni* is the more drought-tolerant taxon that also later moved west, following the westward movement of *Q. parvula*. As the climate continued to dry in the region, *Q. wislizeni* continued to expand westward with *Q. parvula* contracting even farther west, eventually resulting in its current coastal distribution.

One other coastal red oak, *Q. agrifolia*, is well supported except that the southernly distributed var. *oxyadenia* (Torr.) J.T. Howell, while morphologically and elevationally distinct (>750 m), is weakly supported by molecular data [23,95]. In this case, it's possible that the var. *oxyadenia* morphology has arisen multiple times in response to climatological differences in the drier, higher elevation and more southerly environments in which it is found. Future studies should include disjunct Baja populations of *Q. agrifolia* to test the hypothesis that a distinct genetic entity with the morphology of var. *oxyadenia* inhabits the more isolated areas farther south, and that widespread interbreeding between the var. *agrifolia* and var. *oxyadenia* forms has confounded that distinction just north within the FNA region.

4.2. Quercus subsect. Palustres (Trel.) A. Camus, Monogr. Genre Quercus 3: 360. 1952

Large trees to 25 m; **Bark** gray, brown, with flat ridges and shallow to broad fissures; **Twigs** slender, glabrous; **Buds** gray to reddish brown, ovoid 3–7 mm; **Leaves** deciduous, petiole 20–60 mm; **Leaf blade** ovate, elliptic, obovate; base cuneate to truncate; margin with 5–11 lobes, and 9–30 awns, lobes acute to distally expanded, apex acute to acuminate; surfaces abaxially glabrous except for conspicuous axillary tufts of fasciculate trichomes; **Acorns** biennial, cup thin, scale bases visible on inner surface, saucer to deeply goblet shaped with minor to pronounced constriction at the base, scale tips appressed, nut globose to broadly ovoid; cotyledons distinct.

Distribution: Northern and Atlantic States, north to Canada (ONT), west to OK, KS, south to eastern TX, LA (see Figure 2)

Includes: *Q. palustris* Muenchh. (the type), *Q. texana* Buckley

Palustres include two largely allopatric bottomland species with highly similar vegetative features that were recognized by Camus to be distinct among lobe-leaved red oaks within North America. However, both Camus and Trelease thought that *Q. palustris* and *Q. georgiana* were closely related, while our analyses place *Q. georgiana* within subsect. *Phellos*. Although apomorphies are not obvious, phenetic studies using morphological data from leaves and fruit closely clustered *Q. palustris* and *Q. texana* (=*Q. nuttallii* E.J. Palmer) to the exclusion of other lobe-leaved species of *Lobatae* [101].

The first genetic result to suggest that at least *Q. palustris* was divergent from other eastern North American lobe-leaved species was obtained by allozyme analyses [102,103]. Phylogenomic studies confirm the early genetic work and resolve *Palustres* as the first branch among the eastern North American *Lobatae*, sister to the remainder of the section. For the two taxa of *Palustres*, dating analysis also shows a deep split, ca. 20 Ma, a pattern consistent with the branching structure within four of the five early-diverging red oak lineages within the FNA region.

4.3. Quercus subsect. Coccineae *(Trel.) A. Camus, Monogr. Genre* Quercus 3: 386. 1952

Q. subsect. *Velutinae* (Trel.) A. Camus, Monogr. Genre *Quercus* 3: 376. 1952.

Shrubs and medium to large trees to 30 m; **Bark** dark brown to black, gray-brown, furrowed, ridges wide, shiny or with scaly fissures; **Twigs** somewhat thick, glabrous or sparsely pubescent; **Buds** reddish brown, tawny, gray, or silver, conic to ovoid-ellipsoid 3–12 mm; **Leaves** deciduous, petiole 20–60 mm; **Leaf blade** circular, ovate, elliptic, ovate to obovate; base cuneate, obtuse to truncate; margin with 5–11 lobes, and 12–55 awns, lobes acute, ovate-oblong, to distally expanded, apex obtuse, acute, to acuminate; surfaces abaxially glabrous or pubescent, trichomes multiradiate or stipitate fasciculate in conspicuous axillary tufts; **Acorns** biennial, cup thin to thick, cup to saucer-shaped, hemispheric to turbinate, scales smooth to tuberculate, scale tips appressed to loose at the margin, nut ovoid, subglobose to oblong; cotyledons distinct.

Distribution: Northern and Atlantic States, north to Canada (N.B., N.S., ONT, QUE, P.E.I.), west to KS, OK, TX, south to FL (see Figure 2)

Includes: *Q. acerifolia* (Palmer) Stoynoff and Hess, *Q. buckleyi* Nixon and Dorr, *Q. coccinea* Muenchh. (the type), *Q. ellipsoidalis* E.J. Hill, *Q. rubra* L., *Q. shumardii* Buckland, *Q. velutina* Lam.

The circumscription of this group of eastern North American lobe-leaved tree species by Trelease is nearly fully supported by our analyses. The few exceptions include *Q. velutina* which had been either aligned with *Q. marilandica* or treated as a monotypic [72,75,90], and the inclusion of *Q. gravesii* (= *Q. texana* var. *chesosensis* Sarg.) by Trelease, which we resolved as sister lineage to *Erythromexicana* insofar as we have sampled it to date (see above).

Coccineae are defined by a certain phenetic similarity and regional proximity. One distinction shared by many of these species (e.g., *Q. rubra, Q. shumardii*) is wide geographic range, which promotes differentiation in morphological, physiological and ecological traits, and patterns of variation that have been variously treated as species and varieties [104,105]. We resolved two subclades that correspond to the main morphological groups that have been generally recognized: 1) (*Q. acerifolia—Q. buckleyi—Q. shumardii*) + *Q. rubra*); and 2) (*Q. velutina* + *Q. coccinea*) + *Q. ellipsoidalis*).

The first group includes the *Q. shumardii* complex, a cluster of continuously varying morphological forms that have been recognized as varieties and species (e.g., *Q. acerifolia*) [103]. Edaphic specialization appears to be an important driver of differentiation within this complex as *Q. buckleyi* (on limestone in TX and OK), *Q. shumardii* var. *schneckii* (Britton) Sarg. (uplands west of the Blue Ridge), and *Q. acerifolia* (on sandstone, shales) are all endemic to the south-central Midwest. Additional range extensions of *Q. acerifolia*-like populations on limestone into east central Alabama are likely (W. Finch, pers. comm.), and fine-scale molecular and morphological studies of the complex are underway (Y. Wu, pers. comm.). The taxonomic history of *Q. rubra* involves an intraspecific taxon often associated with the higher elevations in the Appalachians, ca. above 1000 m (e.g., var. *ambigua* (A. Gray) Fernald = var. *borealis* (Michx. f.) Farw.). These populations have been quantitatively distinguished by phenetic analysis of morphometric data [101] and by flavonoid profiles [106], but intergradation between the two morphotypes precludes taxonomic recognition [107]. Additional work is needed to better understand the nature of this clinal pattern of variation.

The second group within subsect. *Coccineae* includes two widespread species, *Q. coccinea* and *Q. velutina*, and the north-central endemic *Q. ellipsoidalis* [108,109], which has the distinction of being our northernmost member of sect. *Lobatae* [110,111] and, historically, a locus of substantial taxonomic confusion in the upper Midwest (e.g., [111–117]). The question of whether *Q. ellipsoidalis* is distinct from *Q. coccinea* has been studied with AFLP markers [43,118], microsatellites [59], and RAD-seq data [15] using population-level sampling, including many of the morphologically similar red oaks (including *Q. palustris*) that co-occur in the northern part of the FNA region. There is strong evidence for recognizing *Q. ellipsoidalis* as a distinct species based on this fairly broad sampling. Further, there is little evidence for gene flow between *Q. ellipsoidalis* and *Q. coccinea*, which resolve as largely allopatric sister species, but a low level of ongoing gene flow between *Q. ellipsoidalis* and both *Q. velutina* and *Q. rubra* [43,119–121]. The more distantly related *Q. rubra* and *Q. ellipsoidalis* also show evidence of genomically heterogeneous introgression driven by selection: alleles that distinguish the two species at a gene associated with flowering phenology segregate along moisture gradients and are preferentially shared between species in intermediate environments [122,123]. Genomic study of this species is underway (O. Gailing, pers. comm.). This western Great Lakes species may turn out to be a gateway to broader understanding of speciation in red oaks.

4.4. Quercus subsect. Phellos (G. Don) A. Camus, Monogr. Genre Quercus 3: 293. 1952

Q. subsect. *Myrtifoliae* (Trel.) A. Camus, Monogr. Genre *Quercus* 3: 280. 1952; *Q.* subsect. *Nigrae* (G.Don) A. Camus, Monogr. Genre *Quercus* 3: 322. 1952; *Q.* subsect. *Marilandicae* (Trel.) A. Camus, Monogr. Genre *Quercus* 3: 329. 1952; *Q.* subsect. *Laevis* A. Camus, Monogr. Genre *Quercus* 3: 338. 1952; *Q.* subsect. *Pagodifoliae* (Trel.) A. Camus (as *Pagodaefoliae*), Monogr. Genre *Quercus* 3: 343. 1952; *Q.* subsect. *Ilicifoliae* (Oersted) Trel. A. Camus, Monogr. Genre *Quercus* 3: 353. 1952.

Shrubs or large trees to 40 m; **Bark** hard, black, roughly furrowed separated by smooth furrows or with rectangular blocks; **Twigs** slender to thick, glabrous or glabrescent or pubescent; **Buds** light brown, reddish-brown to purplish, conic to ovoid 2–10 mm; **Leaves** deciduous or subevergreen, petiole 1.5–60 mm; **Leaf blade** elliptic, ovate, obovate, circular to rhombic; base cuneate, rounded or cordate; margin entire, revolute or with 3–11 lobes, and 1–25 awns, lobes acute, attenuate to falcate; surfaces abaxially glabrous to tomentulous to uniformly pubescent, trichomes multiradiate, rosulate or stipitate fasciculate in axillary tufts; **Acorns** biennial and annual, cup thin to involute, shallow goblet-shaped, saucer to cup-shaped, scale tips appressed to loose at the margin, nut ovoid to subglobose; cotyledons distinct.

Distribution: Atlantic and Gulf region States, north to IA, MN, west to OK, TX (see Figure 2)

Includes: Q. arkansana Sarg., * Q. elliottii Wilbur, Q. falcata Michx., Q. georgiana M. A. Curtis, Q. hemisphaerica Bartram ex Willd., * Q. ilicifolia Wangen., Q. imbricaria Michx., Q. incana Bartram, * Q. inopina Ashe, Q. laevis Walter, Q. laurifolia Michx., Q. marilandica Muenchh.,*Q. myrtifolia Willd., Q. nigra L., Q. pagoda Raf., Q. phellos L. (the type)

The center of species diversity for *Phellos* falls within the North Atlantic Coastal Plain (NACP), a recently recognized global biodiversity hotspot [124]. The combined distributions of eleven of the 16 species conform to the broad outline of the NACP, while the ranges of three of the 11 species, *Q. hemisphaerica*, *Q. phellos* and *Q. nigra*, have expanded from the southeastern coastal plain into adjacent piedmont, likely due to a combination of ecological disturbance and escape from cultivation [107].

Resolution of these 16 species aligns a heterogeneous and variously treated set of species into subsect. *Phellos*, making it the largest subsection within the FNA region. It is clear from previous treatments that a subset of this diversity was recognized as a natural group based on sharing the non-lobed laurel-like leaf type, small acorns, cups covering less than half of the nut, with scales closely appressed (subsect. *Phellos* sensu Camus: *Q. imbricaria, Q. incana, Q. laurifolia, Q. phellos, Q. pumila* Walter (= *Q. elliottii*). Other potentially related but phenetically dissimilar taxa with lobed leaves (e.g., *Q. falcata, Q. ilicifolia, Q. marilandica*) instead were segregated into series and subsects, often consisting of one or two species ([72,75]; see [125] for review).

Our results are generally consistent with more recent studies of flavonoid profiles and leaf architecture across these taxa: the systematics of the eight species treated by Trelease and later by Camus within their concepts *Laurifoliae, Marilandicae,* and *Nigrae* cannot be divorced from other taxa within the southeastern FNA region, e.g., *Q. laevis, Q. myrtifolia,* and *Q. inopina* [125]. The suggestion of a hybrid origin for many of these taxa [125] does not appear to be supported by phylogenomic study [15,19], which clearly resolves these species, with the caveat that this radiation arose relatively recently, within the last 10 Ma.

More intensive and targeted sampling is generally needed to test specific cases of reticulate evolution, and this clade has yet to be investigated so closely.

Based on population level sampling, we found that several of the lobe-leaved species show a close relationship to species with unlobed leaves. Additionally, although the five subclades resolved in our analyses are not strongly supported, the following preliminary patterns of species relationship suggest that diversification may be traced along several underlying ecological trajectories. For example, the group formed by (*Q. hemisphaerica—Q. incana—Q. myrtifolia—Q. inopina*) share unlobed leaves and occupy core NACP habitats of mostly sandhills; by contrast, the group formed by (*Q. ilicifolia—Q. georgiana—Q. imbricaria*) tend to occur in upland areas outside of the NACP in various habitats and substrates, including ravines and granitic outcrops, except that Q. *ilicifolia* ranges onto the northern edge of the NACP from New Jersey to Maine. Two other sets of species span sandhills to bottomlands. The larger of the two comprises (*Q. elliottii—Q. pagoda—Q. phellos—Q. laurifolia—Q. nigra—Q. arkansana*), which includes only three taxa that Camus placed with her concept of *Phellos*. A close relationship between *Q. arkansana* and *Q. nigra* is supported here [125]. The final group ((*Q. laevis* + *Q. marilandica*)—*Q. falcata*) tends to occupy sandy well-drained poor soils spanning coastal areas to adjacent uplands. Two of the tree species often resolved as sister taxa, *Q. laevis* and *Q. marilandica,* share a combination of bark, twig, and foliar characteristics. We also sampled *Q. marilandica* var. *ashei* Sudw. from the type location (Scott Mt.) in the Cross Timbers Region of Oklahoma. The placement of two accessions as sister to var. *marilandica* supports previous evidence to recognize this taxon as distinct [88].

5. Section Quercus–(45/ca. 150)

The white oaks stand out as the most species rich and most broadly distributed of all oak clades (Figure 1). The center of species diversity, ca. 100 spp., occurs in Mexico [84], but expansion of white oaks into the driest shrublands of the southwest (e.g., *Q. turbinella*, *Q. cornelius-mulleri*) and Intermountain west (*Q. gambelii*), and their extension into the coldest woodlands of the northern Midwest (*Q. macrocarpa*) demonstrate a physiological breadth not seen in the *Lobatae* [17]. Although white oaks share diagnostic features such as basal abortive ovules, glabrous endocarp, and annual fruit maturation with sects. *Virentes* and *Ponticae*, they are easily distinguished from *Virentes* and to a lesser extent *Ponticae* [3].

As with the *Lobatae*, our analyses reveal that the earliest branch of white oaks, which we treat here as subsect. *Dumosae*, is distributed largely within the CA-FP. It is sister to two major clades: One comprises three clades of mostly north temperate species that range from eastern North America to Eurasia, the other, a succession of three clades: the predominantly eastern subsect. *Stellatae*; a predominantly Texan group we treat here as subsect. *Polymorphae*; and the more recently derived Mexican and Central American clade, informally called *Leucomexicana* pending more detailed study. For the FNA region, we sampled every species except for *Q. intricata* Trel. (Trans-Pecos Region; Chihuahuan Desert Region) and *Q. carmenensis* C.H. Muller (Chisos Mts, Sierra del Carmen Region, Coahuila, Mexico); both of these unsampled species are likely to be placed into the groups we have recognized in this treatment. We also place ten additional FNA species distributed along the southwestern edge of the FNA region into at least three subclades within *Leucomexicana*. These species, a mix of shrubs and trees ranging from open chaparral, woodlands, and slopes of the Chihuahuan Desert Region west to the southern CA-FP, had been placed into many different series and subsections by Trelease and Camus. While additional sampling is needed for formal recognition, eight of these species, including *Q. engelmannii* from the CA-FP [35,126,127] and *Q. turbinella* from the Sonoran Desert, fall into one clade, suggesting a distinct east-west regional radiation that straddles the southwestern border of the FNA region that extends into Mexico.

Where the ranges of distantly related white oaks narrowly intersect, species have formed hybrid zones in areas of sympatry. One case study includes the tree species *Q. engelmannii* and two shrub species, *Q. cornelius-mulleri* and *Q. berberidifolia*, both placed here in subsect. *Dumosae*. Morphological observations of hybridization remain anecdotal, but recent molecular analyses using phylogenetic methods to estimate whether allele sharing is due to gene flow conclude that ancient introgression had shuttled alleles from *Q. cornelius-mulleri* into *Q. engelmannii* [127]. In another case study, *Q. gambelli*, the only white oak of the Intermountain west, and here placed in subsect. *Dumosae* (see below), has long been recognized for hybridizing with white oak species of at least four distinct subclades that we recognize, including *Q. turbinella* and *Q. grisea* of *Leucomexicana* (e.g., [128,129]) and *Q. macrocarpa* of subsect. *Prinoideae* [130]. While recent genetic analyses show that morphologically typical *Q. gambelii* is genetically homogeneous (samples from AZ, CO, NM, UT), there is strong evidence that ancient introgression with *Q. macrocarpa* causes it to be unstable in phylogenomic analyses [25]. The fact that secondary contact zones exist between well separated white oak lineages makes them ideal case studies for ancient introgression and the development of more complete understanding of the nature of the mosaic structure of the oak genome [15,127,131].

Reconstruction of the timing and biogeography of the *Quercus* phylogeny also suggests an initial diversification of white oaks within the FNA region, followed by simultaneous movement south in a pattern closely parallel to *Lobatae*, then across the entire geographic range of Eurasia. The direction of movement, i.e., west through the Beringian Land Bridge or east through the North Atlantic Land Bridges, into Eurasia remains ambiguous, however, recent analyses based on whole plastome genome data also suggest a clear origin of the roburoids within the Americas (B. Wang, P.S. Manos et al. in prep.). The two relatively young and nearly contemporaneous radiations, *Leucomexicana* and roburoids, starting about 15 Ma [15], together account for roughly 75% of the diversity of the section (see Figure 2).

In this subsectional treatment, we re-classify a total of 33 species of sect. *Quercus* recovered in our global analysis, and resolved with high confidence into five clades.

5.1. Quercus subsect. Dumosae *(Trel.) A. Camus, Monogr. Genre* Quercus *2: 462. 1936*

Q. subsect. *Douglasieae* (Trel.) Camus, Monogr. Genre *Quercus* 3: 666. 1936; *Q.* subsect. *Lobatae* (Trel.) Camus, Monogr. Genre *Quercus* 3: 681. 1936; *Q.* subsect. *Gambelieae* (Trel.) Camus, Monogr. Genre *Quercus* 3: 694. 1936.

Shrubs, small to large trees to 30 m; **Bark** light to dark gray, dark brown, scaly or deeply checkered in age; **Twigs** slender, yellowish, gray, brown or reddish brown, densely puberulent to tomentulose with spreading hairs to glabrate; **Buds** yellowish, brown to reddish brown, ovate, globose, ovoid to fusiform, 1–12 mm; **Leaves** deciduous or subevergreen, petiole 1–20 mm; **Leaf blade** broadly obovate, elliptic, subrotund to ovate or oblong to oblanceolate, deeply to shallowly 4–8 lobed; base truncate, rounded-attenuate, cuneate or cordate; margins entire, irregularly shallowly toothed, coarsely toothed to spinose-toothed or lobed, lobes oblong to spatulate, obtuse, rounded, subacute or blunt; surfaces abaxially grayish to whitish, dull green, yellowish to light green, blue-green to glaucous, waxy or glossy, densely velvety with erect 4–8 rayed fasciculate trichomes or dense to sparsely appressed to semi-erect 8–12 rayed stellate trichomes, sometimes with interlocking or fused rays, becoming glabrous; **Acorns** annual, 1–3 subsessile or pedunculate, cup saucer-shaped, hemispheric, deeply cup-shaped or turbinate, base flat or rounded, scales closely appressed, ovate, weakly to strongly tuberculate, scale tips acute, mostly free often reflexed, nut conic, ovoid to globose, ellipsoid, oblong to fusiform, cylindric or barrel-shaped; cotyledons distinct.

Distribution: Pacific States, north to Canada (B.C.), south to Mexico (B.C.) (see Figure 2)

Includes: * *Q. berberidifolia* Liebm., * *Q. cornelius-mulleri* Nixon and K. P. Steele, *Q. douglasii* Hook. and Arn., * *Q. dumosa* Nutt. (the type), * *Q. durata* Jepson, * *Q. gambelii* Nutt., *Q. garryana* Douglas ex Hook, * *Q. john-tuckeri* Nixon and C. H. Muller, *Q. lobata* Née, * *Q. pacifica* Nixon and C. H. Muller

Dumosae form the sister group to all white oaks, diverging by at least 40 Ma likely from within the FNA region. Two clades of basically sclerophyllous shrubby species are nested within an early-branching grade of the two deciduous lobe-leaved species, *Q. lobata* and *Q. garryana* [27]. Fossil leaf compressions attributed to precursors of *Q. lobata* (e.g., *Q. prelobata*) occur well north of the current distribution, indicating a broader ancestral distribution of the stem group going back to the early Miocene, 16–23.5 Ma [99,100].

In addition to the nine *Dumosae* taxa that occur in the Pacific States, we expand the subsection to include *Q. gambelii*, consistent with the high degree of leaf and trichome similarity between *Q. gambelii* and *Q. garryana sensu lato* [132]. *Quercus gambelii* is a curiously widespread species that consists of many isolated but weakly differentiated populations [77]. It was the eponymous center of the 11-species Treleasian series *Gambelieae*, but all those taxa have subsequently been subsumed into *Q. gambelii* or are otherwise not recognized. It was never classified with other lobed-leaved species [72,74,90]. Its uncertain systematic affinities may have been partly a consequence of its isolated geography. In the northern part of its range, it is often the only oak species within the Intermountain west. But in the southwest, *Q. gambelii* is surrounded by and often sympatric with multiple species representing at least four subsects. of white oaks, resulting in a rich taxonomic and ecological literature describing patterns of natural hybridization [128,129,133,134].

However, its uncertain position may be due to its cryptic morphology, a lobedness that may derive in part from an ancient history of introgression with eastern lobed white oaks. Our phylogenomic studies detected a primary signal of phylogenetic history between *Q. gambelii* and *Q. lobata* as the branches leading to accessions of both species formed a clade [25]. The position of this clade was determined to be unstable in the phylogenetic space between *Dumosae* and *Prinoideae*. Our work, based on tests designed to detect secondary phylogenetic signals due to introgression, suggests that ancient hybridization between *Q. gambelii* and *Q. macrocarpa* is the likely cause of phylogenetic uncertainty in the placement of this species. Our inference of historical rather than contemporary gene flow is further supported by the fact that *Q. gambelii*, *Q. macrocarpa*, and *Q. lobata* are strongly monophyletic across analyses. If ongoing gene flow were the cause of phylogenetic instability, we would expect some individuals of these species to switch positions between species or become dragged to phylogenetically intermediate positions themselves. Instead, all individuals of these species form separate clades by species.

Thus *Q. gambelii* seems to have arisen from allopatric speciation within the formerly widespread range of *Q. lobata* or the ancestral complex formed by *Q. lobata* and the named varieties of *Q. garryana* [20,25]. In this scenario, uplift of the Sierra Nevada, increasing aridity, and other developing north-south barriers isolated populations of ancestral *Q. gambelii* during the Miocene. The aridity of the Great Basin would have forced the nascent species eastward, into mesic woodlands, and to the southeast, along the foothills of the Rocky Mountains into its current elevational range of 1000 to 3000 m. While *Q. gambelii* does not currently overlap in range with *Q. lobata*, *Q. garryana*, or *Q. macrocarpa*, periods of range expansion during warm and wet periods over the last 5 million years likely generated extensive secondary contact zones between *Q. gambelii* and *Q. macrocarpa* [99]. In addition to our genetic results, support for historical range extension and hybridization comes from morphological evidence observed in outlier populations of *Q. gambelii* in the Black Hills of South Dakota and *Q. macrocarpa* in northeastern New Mexico [130].

The main diversification within *Dumosae* involves an array of mostly shrub species, including *Q. pacifica*, the second taxon endemic to the California Channel Islands, that have vexed systematists for well over a century [20,132,135]. It was suggested that all of the California shrub species, except for *Q. turbinella* are derived from a lobe-leaved ancestor similar to *Q. lobata*. Nixon [122] outlined anatomical evidence for this relationship, noting the venation pattern shared between sclerophyllous shrub species and lobe-leaved species, specifically branched secondary veins, multiple and irregularly spaced teeth, and an often coarse pattern of lobation. This pattern contrasts with the condition of secondary veins that rarely branch, generally resulting in one tooth per secondary vein and regularly spaced teeth, like those of *Q. turbinella* and other shrub species in the southwestern FNA region and Mexico. Our analyses confirm this hypothesis, while in-depth studies have generated additional patterns of species relationships within the shrub diversity of *Dumosae*.

By ca. 15 Ma, the nested crown lineage of *Dumosae* diverged into two clades, one clade comprising *Q. pacifica*—*Q. dumosa* and a distinct well supported subclade of (*Q. cornelius-mulleri*- (*Q. john-tuckeri* + *Q. douglasii*)), and the other clade formed by a grade of population samples representing the widespread *Q. berberidifolia*, within which are nested two separate subclades of *Q. durata* var. *durata* and var. *gabrielensis* Nixon and C.H. Muller [27,35,127]. This overall pattern and taxonomy are consistent with previous morphological work that segregated the narrow endemic *Q. dumosa* from *Q. berberidifolia* and recognized the parapatrically distributed *Q. cornelius-mulleri* and *Q. john-tuckeri* to be distinct from previous taxonomic treatments (see [132] for review). The nested position of the tree species *Q. douglasii* within the shrub clade formed by *Q. cornelius-mulleri* and *Q. john-tuckeri* is an interesting example of an evolutionary reversal back to the tree habit. The timing of divergence of the nested crown clade of shrubby species, from *Q. pacifica* through the *Q. berberidifolia* clade, is consistent with the hypothesis of Kim et al. (2018) [127]: Diversification of the shrub species coincides with mid-Miocene development of the Mediterranean climate in the CA-FP [136]. The onset of this climate regime would have increased the frequency of fire, potentially favoring shrubs over trees. The evolution of *Q. douglasii*, a tree with shrub-like traits, may have been enabled by ecological opportunity in the relatively mesic foothills of the Sierra Nevada.

Recent demographic studies of gene flow among different combinations of shrub species suggest that hybridization continues to promote taxonomic uncertainty while ecology and climate reinforce genomic cohesiveness and identity [35,127,137]. The full complement of genetic analyses confirms that most of the morphologically-defined species of *Dumosae* are cohesive and often ecologically isolated to some degree. One notable exception stands out from studies of the *Q. berberidifolia*—*Q. durata* complex, which have revealed the most challenging case to date. *Quercus durata* appears to have arisen twice from within *Q. berberidifolia*, despite the generally clear taxonomic distinctions between *Q. berberidifolia* and *Q. durata*. While *Q durata* var. *durata* is well separated, albeit nested, within *Q. berberidifolia*, var. *gabrielensis* is only weakly distinguished [27,127]. This distinction likely involves ecological isolation, as *Q. durata* var. *durata* mostly occurs on serpentine soils. Such cases may be more widespread in oaks than we suspect and bear closer investigation.

5.2. Quercus subsect. Prinoideae (Trel.) A. Camus, Monogr. Genre Quercus 2: 434. 1936

Q. subsect. *Lyratae* (Oersted) A. Camus, Monogr. Genre *Quercus* 2: 723. 1936; *Q.* subsect. *Macrocarpae* (Trel.) A. Camus, Monogr. Genre *Quercus* 2: 738. 1936.

Shrubs or large trees to 30–50 m; **Bark** dark gray to light gray, thin flaky to papery, scaly, flat-ridged or with thick plates underlying scales; **Twigs** coarse, slender to moderate, gray, tan, reddish brown, fine-pubescent to villous becoming glabrate or with round to radiating flat corky wings; **Buds** gray, brown, red-brown, subrotund to broadly ovoid 1–4 mm; **Leaves** deciduous, deciduous or persistent stipules, petiole 4–30 mm; **Leaf blade** moderate to large, lanceolate, obovate to narrowly elliptic or broadly obovate; base rounded, truncate, cuneate to acute; margins regularly undulate, toothed or shallow-lobed to deeply lobed; surfaces abaxially glaucous to light green, tomentose, trichomes erect long-rayed fasciculate and small stellate; **Acorns** annual, 1–3 on thin to stout peduncle, cup hemispheric, cup-shaped, turbinate to goblet-shaped or spheroid, base rounded, scales closely appressed to laterally connate, moderately to prominently keeled, tuberculate or coarsely thickened and fringed at the margin with short awns, nut ovoid-ellipsoid or oblong; cotyledons distinct.

Distribution: Atlantic States and Mississippi Valley, north to Canada (MAN, N.B., ONT, QUE, SASK), south to the mountains of western TX, NM, and northeastern Mexico (COAH, N.L., HGO, TAMPS) (see Figure 2)

Includes: *Q. bicolor* Willd, *Q. lyrata* Walter, *Q. macrocarpa* Michx., *Q. muehlenbergii* Engelm., * *Q. prinomides* Willd. (the type)

Prinoideae form the sister to a moderately sized temperate clade that spans eastern North America and Eurasia. The five species we recognize here share a common set of trichomes and similar cup scale morphology. Camus classified *Q. montana* and *Q. michauxii* in her concept of subsect. *Prinoideae*, while recognizing *Q. lyrata* and *Q. macrocarpa* in separate monotypic groups. Our analyses placed the former species pair with *Q. alba*, consistent with their lack of the trichome types observed across all *Prinoideae* [138,139].

Extensive population sampling and analysis resolved a deep split within the crown group, ca. 20 Ma, pairing tree species *Q. muehlenbergii* and shrub *Q. prinoides* sister to a well-supported group of (*Q. macrocarpa-* (*Q. lyrata* + *Q. bicolor*)). As with morphology, our data are unable to detect a clean separation between the widespread and ecologically distinct "species-pair," *Q. muehlenbergii* (limestone) and *Q. prinoides* (sands, shales). Further studies are needed to understand the ecological processes behind the weak genetic separation of taxa, despite the heritable differences observed in *Q. prinoides*, such as the shrub habit and early maturation to flower and fruit [132]. It may be that the situation in *Q. muehlenbergii* mirrors that of *Q. berberidifolia* and *Q. durata*, in which the same phenotype has arisen more than once due to parallel selective processes across the range of the more widespread species.

5.3. Quercus subsect. Albae (G. Don) A. Camus, Monogr. Genre Quercus 2: 728. 1936

Large trees to 30 m; **Bark** dark gray or brown with V-shaped furrows or light brown to gray, scaly; **Twigs** moderate, green, brown to reddish brown, glabrous or fine-pubescent to sparse spreading hairs, becoming glabrous; **Buds** gray, light brown, reddish-brown, ovoid 3–6 mm; **Leaves** deciduous, petiole 5–30 mm; **Leaf blade** moderate, broadly obovate to narrowly elliptic or narrowly obovate; base cuneate to acute or rounded acuminate to broadly cuneate; margins regularly toothed or moderately to deeply lobed; surfaces abaxially light green or yellowish; trichomes solitary, appressed, and erect or appressed, 1–2 rayed fasciculate, persistent to quickly shed; **Acorns** annual, 1–3 sessile or pedunculate, cup shallowly cup-shaped to hemispheric or deeply goblet shaped, base rounded, rim thin to moderate, scales concentric to laterally connate or loosely to closely appressed, slightly keeled to prominently tuberculate, nut ovoid-ellipsoid, cylindric or oblong; cotyledons distinct.

Distribution: Atlantic States and Mississippi Valley, north to Canada (ONT, QUE), south to eastern TX, northern FL (see Figure 2)

Includes: *Q. alba* L. (the type), *Q. michauxii* Nutt., *Q. montana* Willd.

The three *Albae* species are quite divergent with respect to morphology, especially features of the bark and cup scales. Although the leaves of *Q. alba* are generally described as glabrous and frequently are at maturity [130], fasciculate trichomes are present in the early stages of leaf development, and simple trichomes may persist as the leaf expands. *Quercus michauxii* and *Q. montana* also share fasciculate trichomes, but these are persistent, and erect with long rays in the former and appressed with short rays in the latter. *Albae*, as delimited here, lack the stellate trichomes observed in *Prinoideae* [138,139]. Our analyses break-up the so-called chestnut group of the FNA region (*Q. michauxii, Q. montana, Q. muehlenbergii, Q. prinoides, Q. sadleriana*), leading to the distribution species with the 'prinoid' or 'castaneoid' leaf shape into three distinct temperate clades of white oaks: *Albae, Ponticae,* and *Prinoideae*.

Albae combine morphological elements of the prinoids and roburoids to form one of the most genetically cohesive oak subclades of the FNA region. The relationships among the three species (*Q. montana*—(*Q. michauxii* + *Q. alba*)) are well supported and based on strong sampling. The sister group relationship of *Albae* to the roburoids extends the signature pattern of early-diverging deciduous shallowly and deeply lobe-leaved species from the FNA region to Eurasia. This biogeographic connection has been difficult to estimate phylogenetically as a result of ancient hybridization between the *Q. pontica* sublineage of *Ponticae* and ancestral roburoids in Eurasia [24].

5.4. Quercus subsect. Stellatae *(Trel.) A. Camus, Monogr. Genre* Quercus *2: 710. 1936*

Q. subsect *Confusae* (Trel.) A. Camus, Monogr. Genre *Quercus* 2: 671. 1936; *Q.* subsect *Durandieae* (Trel.) A. Camus, Monogr. Genre *Quercus* 2: 676. 1936.

Shrubs, often rhizomatous, moderate to large trees to 30 m; **Bark** light gray or whitish, light brown, scaly or flaky to papery and exfoliating; **Twigs** slender to moderately thick,, densely tomentose, sparsely pubescent, glabrate; **Buds** gray to reddish-brown, ovoid to globose to acute at the apex, 1–6 mm; **Leaves** deciduous to wintergreen, petiole 1–15 mm; **Leaf blade** small to moderately large, obovate to narrowly elliptic, oblong to oblanceolate or rounded 3-dentate to obtriangular; base rounded-attenuate, acute to cordate, cuneate or obtuse; margins undulated, revolute, entire or shallow, moderate to deeply lobed, 3–5 lobes rounded, spatulate or sinuate; surfaces abaxially silvery, grayish, dull to yellowish green or dark green, trichomes dense to scattered, 3–10-rayed stipitate fasciculate, stellate, often becoming glabrate; **Acorns** annual, 1–3 subsessile or pedunculate, cup thin, deeply cup-shaped, goblet-shaped, turbinate, hemispheric, base rounded or constricted, cup scales gray to brown, thin, tightly appressed, flattened to only slightly tuberculate, scale tips acute, nut ovoid, globose, barrel-shaped to elliptic; cotyledons distinct.

Distribution: Atlantic States west to eastern NM, north to southern IA, southeastern NY (see Figure 2).

Includes: *Q. austrina* Small, * *Q. boyntonii* Beadle, * *Q. chapmanii* Sarg., * *Q. havardii* Rydberg, *Q. margarettae* Ashe, *Q. oglethorpensis* W.H. Duncan, *Q. similis* Ashe, * *Q. sinuata* Walter, *Q. stellata* Wangenh. (the type)

Stellatae comprise a balanced mix of trees of various size and low-growing shrubs that occupy mainly lowlands including deep sands, alluvial flatwoods, and bottomlands. Species of this subsection are more or less distributed in the southeast of the FNA region, strongly overlapping in range and habitat type with species of the red oak clade *Phellos*. The placement of *Q. havardii* within this subsection expands the range farther west to eastern New Mexico.

In addition to the five species Camus recognized in her subsect. *Stellatae* (*Q. boyntonii*, *Q. chapmanii*, *Q. margarettae*, *Q. similis*, *Q. stellata*), our expanded delimitation of the group to nine species includes three species—*Q. austrina*, *Q. havardii* [66], and *Q. sinuata*—that had been treated in small subsections of one or two species each, and the unclassified and most recently described oak species east of the Mississippi, *Q. oglethorpensis*. Duncan (1950) [140] had suggested a close affinity of *Q. oglethorpensis* to many of the species placed here based on similarities in cup scales, bark, and leaf epidermal features. Strong similarities in cup shape and well-defined cup scales that are thin, generally narrow, and flat and only slightly tuberculate also support the expansion of the historically recognized post oak subgroup sensu Camus to include four additional species.

Stellatae diversified within the last 15 Ma, around the same time as subsect. *Phellos* and clades within *Leucomexicana*. We resolved *Q. sinuata* and *Q. oglethorpensis* as early-branching species, consistent with Duncan's (1950) view that both are somewhat isolated evolutionarily. The remaining seven species are resolved into three moderately supported subclades: *Q. margarettae*—*Q. austrina*; *Q. chapmanii*—*Q. havardii*; and *Q. stellata*, *Q. similis*, and *Q. boyntonii*. Additional population sampling, especially in the southern Gulf states and west into Texas, is needed to better understand the relationship of geography and potential hybridization to this initial estimate of phylogeny.

5.5. Quercus subsect. Polymorphae *(Trel.) A. Camus, Monogr. Genre* Quercus *2: 561. 1936*

Shrubs, often rhizomatous, small to moderate trees to 20 m; **Bark** light gray to brown, papery, scaly or deeply furrowed, sometimes exfoliating in long strips; **Twigs** slender, gray, yellowish to light brown, reddish brown, pubescent to tomentose or soon glabrate; **Buds** gray, brown, reddish-brown, round to ovoid or with apex acute, 0.5–10 mm; **Leaves** subevergreen, deciduous, petiole 2–25 mm; **Leaf blade** small to moderate, elliptic, narrowly lanceolate to oblong, ovate, obovate, subrotund or rotund; base rounded, cordate, cordulate or cuneate; margins thick or thin, flat, revolute to regularly undulate-crisped, entire to toothed, shallowly lobed sometimes with 2–3 spinescent teeth on each side; surfaces abaxially light green, blue-green, dark green, glaucous or whitish, glabrous, floccose or tomentose, trichomes dense to scattered, flat, curly or stiffened stellate, with erect or appressed simple hairs, sometimes becoming glabrous; **Acorns** annual, 1–2 subsessile or pedunculate, cup thin, shallowly to deeply cup-shaped, turbinate, hemispheric or funnel-shaped, cup scales appressed, thickened basally, moderately to strongly tuberculate, scale tips appressed; nut ovoid-ellipsoid, oblong or barrel-shaped, sometimes flattened at both ends; cotyledons distinct or connate.

Distribution: Southwestern States AZ, NM, OK, TX, northeastern Mexico south to Guatemala (see Figure 2)

Includes: * *Q. hinckleyi* C.H. Muller, *Q. laceyi* Small, *Q. mohriana* Buckley, *Q. polymorpha* Schltdl. and Cham. (the type), *Q. pungens* Liebm., *Q. vaseyana* Buckley

Polymorphae form a small and variable clade that diverged from *Leucomexicana* more than 15 Ma. Of the six recognized taxa, only *Q. hinckleyi* and *Q. polymorpha* share the same Bioregion climate cluster with *Leucomexicana* taxa based on climatic ranges estimated from Worldclim data [77], while the other four species cluster with Eastern North American taxa [19]. The clade is remarkably heterogeneous in morphology and habit, with a patchy distribution in the Chihuahuan Desert Region and adjacent Edwards Plateau except for *Q. polymorpha*, which ranges from southwest Texas, south along the east slope of the Sierra Madre Oriental to Guatemala [83,132].

One well supported subgroup of taxa comprises *Q. hinckleyi*—*Q. pungens*—*Q. vaseyana*. *Quercus pungens* and *Q. vaseyana* have been linked taxonomically, the latter initially treated as a variety of the former, and both classified by Muller (1951) [90] in his series *Vaseyanae*. Muller described and placed *Q. hinckleyi* within series *Glaucoideae*, a group he delimited to include the FNA taxa *Q. laceyi* and *Q. depressipes* Trel. and the widespread Mexican species *Q. glaucoides* M. Martins and Galeotti. Our analyses instead support a regional radiation of taxa, including *Q. laceyi*, while placing the FNA taxa associated with the historical concept of series *Glaucoideae* (e.g., *Q. depressipes*, *Q. engelmannii* Greene, *Q. arizonica* Sarg.) within *Leucomexicana* (Figure 2).

Except for *Q. pungens* and *Q. vaseyana*, none of the taxa treated here within *Polymorphae* have been considered closely related in previous classifications. For example, *Q. laceyi, Q. polymorpha*, and *Q. mohriana* were placed in different sets of mostly Mexican taxa: (1) *Q. mohriana* with *Q. arizonica* in subsect. *Arizonicae* sensu Camus [74]; (2) *Q. polymorpha* with *Q. porphyrogenita* Trel. and a few other Mexican taxa of doubtful validity within subsect. *Polymorphae* sensu Camus [83]; and (3) *Q. laceyi* within series *Glaucoideae* sensu Muller, and as noted later [141], with putative affinities to either *Q. porphyrogenita* or eastern North American white oaks.

Our current sampling and high level of resolution and support within section *Quercus* are sufficient to conclude that most of the previous speculation on the disparate affinities of these taxa is not supported. We sampled two accessions of the widespread *Q. polymorpha*, one from the northernmost population in Texas (Dolan Falls, Val Verde Co.) the other from south of Monterrey (Nuevo Leon; Mexico), about 600 km apart, and they fell sister to one another with strong support [15]. Similarly, each taxon with the exception of *Q. hinckleyi* has been sampled from two separate populations, with all individuals of these species forming tight, cohesive clades.

While our sampling in Mexico is preliminary, analyses suggest at least one compelling scenario for the biogeographic history of this clade. Aside from *Q. polymorpha*, these taxa currently form the most narrowly distributed radiation of the North American oak clades, and the timing of the clade's crown diversification coincides closely with the mid-Miocene expansion of arid zones [142]. Species of this clade generally occur on limestone and desert slopes of the southwestern states (AZ, NM, TX, OK) and three north-central states of Mexico (CHIH, COAH, N.L.). The more mesophytic and broadly distributed *Q. polymorpha* sublineage occurs to the south along the east slope of the Sierra Madre Oriental. Current patterns of distribution suggest a concentrated distribution of relictual species of various habit and climatic niche breadth, with a notable biogeographic disjunction between the *Q. polymorpha* and *Q. mohriana* sublineages by the end of the Pliocene, ca. 3–5 Ma, well after the diversification of main radiation of *Leucomexicanae* [19]. It is likely that the historical range of *Polymorphae* was more widespread farther north and west into the FNA region and south into Mexico during wetter and more mild climates during the late Wisconsin glaciation episode (22–11 thousand years ago). The most distinctive of these taxa, *Q. hinckleyi*, is a narrow endemic of the Trans-Pecos Region, with spinescent, subrotund leaves. It has been called a "climate relict" based on leaf fossil evidence from packrat middens throughout the Pleistocene ([143], for discussion see [144]).

6. Conclusions

We have provided a review and synthesis of the systematics of the oaks within the Flora of the North American region, as informed by recent analyses using phylogenomic analyses of nuclear sequence data. A new classification for part of the American oak clade is presented and discussed with reference to biogeographic history, species-level relationships, and previous hypotheses of hybridization. This update on the evolutionary diversification of an ecologically and economically important tree group provides both a capstone and roadmap to guide future studies.

The recognition of oak clades below the sectional level provides an opportunity to revisit some of the most interesting topics in ecology, biogeography, and species biology. From the ecological processes involved in community assembly to the adaptive role of hybridization, species-level phylogenies derived from next-generation DNA sequence data will enhance scientific inquiry and promote targeted hypothesis testing. In practical terms, translating phylogenies into informative classification systems improves our ability to communicate with precision about the independent evolutionary radiations of species across the tree of life.

Author Contributions: P.S.M. and A.L.H. contributed equally to all aspects of the manuscript. Both authors have read and agreed to the published version of the manuscript.

Funding: This research was funded by National Science Foundation (DEB-1146488, DEB-1146102) and the Duke University Arts and Sciences Council.

Institutional Review Board Statement: Not applicable.

Informed Consent Statement: Not applicable.

Data Availability Statement: Data and scripts to generate the phylogeny and range maps presented in this paper (Figure 2) are available in two public repositories: GitHub (https://github.com/andrew-hipp/2021-oak-classification, accessed on 16 June 2021), with the published data and scripts released as version 0.95-1 under Zenodo DOI 10.5281/zenodo.4905230 (https://zenodo.org/record/4905230, accessed on 16 June 2021).

Acknowledgments: The authors thanks Alan Whittemore and Al Keuter for advice on the taxonomic treatment and comments on the *Agrifoliae*, respectively. They also acknowledge the constructive comments of two anonymous reviewers and thank Mary Ashley and Janet Backs for the invitation to contribute to this special volume.

Conflicts of Interest: The authors declare no conflict of interest.

References

1. Nixon, K.C. Quercus. In *Flora of North America North of Mexico*; Committee Flora of North America Editorial, Ed.; Oxford University Press: New York, NY, USA, 1997; Volume 3, pp. 445–447.
2. Manos, P.S.; Stanford, A.M. The Historical Biogeography of Fagaceae: Tracking the Tertiary History of Temperate and Subtropical Forests of the Northern Hemisphere. *Int. J. Plant Sci.* **2001**, *162*, S77–S93. [CrossRef]
3. Denk, T.; Grimm, G.W.; Manos, P.S.; Deng, M.; Hipp, A.L. An updated infrageneric classification of the oaks: Review of previous taxonomic schemes and synthesis of evolutionary patterns. In *Oaks Physiological Ecology. Exploring the Functional Diversity of Genus Quercus L.*; Gil-Pelegrín, E., Peguero-Pina, J.J., Sancho-Knapik, D., Eds.; Tree Physiology; Springer: Cham, Switzerland, 2017; pp. 13–38. ISBN 9783319690988.

4.   Cavender-Bares, J. Diversity, Distribution, and Ecosystem Services of the North American Oaks. *Int. Oaks* **2016**, *27*, 37–48.
5.   Fahey, R.T.; Darling, L.; Anderson, J. Oak Ecosystem Recovery Plan: Sustaining Oaks in the Chicago Wilderness Region; Chicago Wilderness Oak Ecosystem Recovery Working Group; USDA Forest Service and US Fish & Wildlife Service, Chicago, Illinois. 2015. Available online: http://chicagorti.org/sites/chicagorti/files/OERP-Full-Report-lowres.pdf (accessed on 12 June 2021).
6.   Nowak, D.J.; Hoehn, R.E.I.; Bodine, A.R.; Crane, D.E.; Dwyer, J.F.; Bonnewell, V.; Watson, G. *Urban Trees and Forests of the Chicago Region*; U.S. Department of Agriculture, Forest Service, Northern Research Station: Newtown Square, PA, USA, 2013.
7.   Chassé, B. Eating Acorns: What Story Do the Distant, Far, and near Past Tell Us, and Why? *Int. Oaks: J. Int. Oak Soc.* **2016**, *27*, 107–135.
8.   Logan, W.B. *Oak: Frame of Civilization*; W.W. Norton & Company Inc.: New York, NY, USA, 2005.
9.   Leroy, T.; Plomion, C.; Kremer, A. Oak Symbolism in the Light of Genomics. *New Phytol.* **2020**, *226*, 1012–1017. [CrossRef]
10.   Mabey, R. *The Cabaret of Plants: Forty Thousand Years of Plant Life and the Human Imagination*, 1st ed.; W.W. Norton & Company Inc.: New York, NY, USA, 2017; ISBN 9780393353860.
11.   Forman, L.L. On the Evolution of Cupules in the Fagaceae. *Kew Bull.* **1966**, *18*, 385. [CrossRef]
12.   Nixon, K.C.; Crepet, W.L. TRIGONOBALANUS (FAGACEAE): TAXONOMIC STATUS AND PHYLOGENETIC RELATIONSHIPS. *Am. J. Bot.* **1989**, *76*, 828–841. [CrossRef]
13.   Manos, P.S.; Doyle, J.J.; Nixon, K.C. Phylogeny, Biogeography, and Processes of Molecular Differentiation in *Quercus* Subgenus *Quercus* (Fagaceae). *Mol. Phylogenet. Evol.* **1999**, *12*, 333–349. [CrossRef]
14.   Oh, S.-H.; Manos, P.S. Molecular Phylogenetics and Cupule Evolution in Fagacee as Inferred from Nuclear CRABS CLAW Sequences. *Taxon* **2008**, *57*, 434–451.
15.   Hipp, A.L.; Manos, P.S.; Hahn, M.; Avishai, M.; Bodénès, C.; Cavender-Bares, J.; Crowl, A.A.; Deng, M.; Denk, T.; Fitz-Gibbon, S.; et al. Genomic Landscape of the Global Oak Phylogeny. *New Phytol.* **2020**, *226*, 1198–1212. [CrossRef] [PubMed]
16.   Denk, T.; Grimm, G.W. Significance of Pollen Characteristics for Infrageneric Classification and Phylogeny in *Quercus* (Fagaceae). *Int. J. Plant Sci.* **2009**, *170*, 926–940. [CrossRef]
17.   Cavender-Bares, J. Diversification, Adaptation, and Community Assembly of the American Oaks (*Quercus*), a Model Clade for Integrating Ecology and Evolution. *New Phytol.* **2019**, *221*, 669–692. [CrossRef] [PubMed]
18.   Cavender-Bares, J.; Kothari, S.; Meireles, J.E.; Kaproth, M.A.; Manos, P.S.; Hipp, A.L. The Role of Diversification in Community Assembly of the Oaks (*Quercus* L.) across the Continental U.S. *Am. J. Bot.* **2018**, *105*, 565–586. [CrossRef] [PubMed]
19.   Hipp, A.L.; Manos, P.S.; González-Rodríguez, A.; Hahn, M.; Kaproth, M.; McVay, J.D.; Avalos, S.V.; Cavender-Bares, J. Sympatric Parallel Diversification of Major Oak Clades in the Americas and the Origins of Mexican Species Diversity. *New Phytol.* **2018**, *217*, 439–452. [CrossRef] [PubMed]
20.   Nixon, K.C. The Oak (Quercus) Biodiversity of California and Adjacent Regions. *Gen. Tech. Rep. USDA For. Serv.* **2002**, *PSW-GTR-184*, 3–20.
21.   González-Rodríguez, A.; García-Oliva, F.; Tapie-Torres, Y.; Morón-Cruz, A.; Chávez-Vergara, B.; Baca-Patiño, B.; Cuevas-Reyes, P. Oak Community Diversity Affects Nitrogen in Litter and Soil. *Int. Oaks J. Int. Oak Soc.* **2019**, *30*, 125–130.
22.   Cavender-Bares, J.; Ackerly, D.D.; Baum, D.A.; Bazzaz, F.A. Phylogenetic Overdispersion in Floridian Oak Communities. *Am. Nat.* **2004**, *163*, 823–843. [CrossRef]
23.   Hauser, D.A.; Keuter, A.; McVay, J.D.; HIpp, A.L.; Manos, P.S. The Evolution and Diversification of the Red Oaks of the California Floristic Province (*Quercus* Section *Lobatae*, Series *Agrifoliae*). *Am. J. Bot.* **2017**, *104*, 1581–1595. [CrossRef] [PubMed]
24.   McVay, J.D.; Hipp, A.L.; Manos, P.S. A Genetic Legacy of Introgression Confounds Phylogeny and Biogeography in Oaks. *Proc. Roy. Soc. B.* **2017**, *284*, 20170300. [CrossRef]
25.   McVay, J.D.; Hauser, D.; Hipp, A.L.; Manos, P.S. Phylogenomics Reveals a Complex Evolutionary History of Lobed-Leaf White Oaks in Western North America. *Genome* **2017**, *60*, 733–742. [CrossRef]
26.   Cavender-Bares, J.; Gonzalez-Rodriguez, A.; Eaton, D.A.R.; Hipp, A.L.; Beulke, A.; Manos, P.S. Phylogeny and Biogeography of the American Live Oaks (*Quercus* Subsection *Virentes*): A Genomic and Population Genetics Approach. *Mol. Ecol.* **2015**, *24*, 3668–3687. [CrossRef] [PubMed]
27.   Fitz-Gibbon, S.; Hipp, A.L.; Pham, K.K.; Manos, P.S.; Sork, V. Phylogenomic Inferences from Reference-Mapped and de Novo Assembled Short-Read Sequence Data Using RADseq Sequencing of California White Oaks (*Quercus* Subgenus *Quercus*). *Genome* **2017**, *60*, 743–755. [CrossRef] [PubMed]
28.   Crowl, A.A.; Bruno, E.; Hipp, A.L.; Manos, P.S. Revisiting the Mystery of the Bartram Oak. *Arnoldia* **2020**, *77*, 6–11.
29.   McIntyre, D.J. Pollen and Spore Flora of an Eocene Forest, Eastern Axel Heiberg Island. *N.W.T. Geol. Surv. Can. Bull.* **1991**, *403*, 83–97.
30.   Liu, Z.; Pagani, M.; Zinniker, D.; DeConto, R.; Huber, M.; Brinkhuis, H.; Shah, S.R.; Leckie, R.M.; Pearson, A. Global Cooling During the Eocene-Oligocene Climate Transition. *Science* **2009**, *323*, 1187–1190. [CrossRef] [PubMed]
31.   Little, E.L. *Atlas of United States Trees, Volume 1, Conifers and Important Hardwoods*; Miscellaneous publication; U.S. Dept. of Agriculture, Forest Service: Washington, DC, USA, 1971.
32.   Little, E.L. *Atlas of United States Trees, Volume 4, Minor Eastern Hardwoods*; Miscellaneous publication; U.S. Dept. of Agriculture, Forest Service: Washington, DC, USA, 1977.
33.   Pham, K.K.; Hipp, A.L.; Manos, P.S.; Cronn, R.C. A Time and a Place for Everything: Phylogenetic History and Geography as Joint Predictors of Oak Plastome Phylogeny. *Genome* **2017**, *60*, 720–732. [CrossRef]

34. Muller, C.H. The Live Oaks of the Series Virentes. *Am. Midl. Nat.* **1961**, *65*, 17–39. [CrossRef]
35. Sork, V.L.; Riordan, E.; Gugger, P.F.; Fitz-Gibbon, S.; Wei, X.; Ortego, J. Phylogeny and Introgression of California Scrub White Oaks (Quercus Section Quercus). *Int. Oaks* **2016**, *27*, 61–74.
36. Eaton, D.A.R.; Hipp, A.L.; González-Rodríguez, A.; Cavender-Bares, J. Historical Introgression among the American Live Oaks and the Comparative Nature of Tests for Introgression. *Evolution* **2015**, *69*, 2587–2601. [CrossRef] [PubMed]
37. Crowl, A.A.; Manos, P.S.; McVay, J.D.; Lemmon, A.R.; Lemmon, E.M.; Hipp, A.L. Uncovering the Genomic Signature of Ancient Introgression between White Oak Lineages (*Quercus*). *New Phytol.* **2020**, *226*, 1158–1170. [CrossRef]
38. Muir, G.; Fleming, C.C.; Schlötterer, C. Species Status of Hybridizing Oaks. *Nature* **2000**, *405*, 1016. [CrossRef]
39. Pearse, I.S.; Hipp, A.L. Phylogenetic and Trait Similarity to a Native Species Predict Herbivory on Non-Native Oaks. *Proc. Natl. Acad. Sci. USA* **2009**, *106*, 18097–18102. [CrossRef]
40. Hipp, A.L. Should Hybridization Make Us Skeptical of the Oak Phylogeny? *Int. Oak J.* **2015**, *26*, 9–18.
41. Hipp, A.L. Oak Research in 2015: A Snapshot from the IOS Conference. *Int. Oak J.* **2016**, *27*, 15–22.
42. Craft, K.J.; Ashley, M.V.; Koenig, W.D. Limited Hybridization between *Quercus Lobata* and *Quercus Douglasii* (Fagaceae) in a Mixed Stand in Central Coastal California. *Am. J. Bot.* **2002**, *89*, 1792–1798. [CrossRef] [PubMed]
43. Hipp, A.L.; Weber, J.A. Taxonomy of Hill's Oak (*Quercus Ellipsoidalis*: Fagaceae): Evidence from AFLP Data. *Syst. Bot.* **2008**, *33*, 148–158. [CrossRef]
44. Lepais, O.; Petit, R.J.; Guichoux, E.; Lavabre, J.E.; Alberto, F.; Kremer, A.; Gerber, S. Species Relative Abundance and Direction of Introgression in Oaks. *Mol. Ecol.* **2009**, *18*, 2228–2242. [CrossRef] [PubMed]
45. Curtu, A.L.; Gailing, O.; Leinemann, L.; Finkeldey, R. Genetic Variation and Differentiation within a Natural Community of Five Oak Species (*Quercus* Spp.). *Plant Biol.* **2007**, *9*, 116–126. [CrossRef] [PubMed]
46. Gailing, O.; Curtu, A.L. Interspecific Gene Flow and Maintenance of Species Integrity in Oaks. *Ann. For. Res.* **2014**, *57*, 5–18. [CrossRef]
47. Dodd, R.; Papper, P. Disentangling the Phylogenetic Network of the California Red Oaks. *Int. Oaks* **2019**, *30*, 203–208.
48. Valencia-A, S. Species Delimitation in the Genus Quercus (Fagaceae). *Bot. Sci.* **2021**, *99*, 1–12. [CrossRef]
49. Hipp, A.L.; Whittemore, A.T.; Garner, M.; Hahn, M.; Fitzek, E.; Guichoux, E.; Cavender-Bares, J.; Gugger, P.F.; Manos, P.S.; Pearse, I.S.; et al. Genomic Identity of White Oak Species in an Eastern North American Syngameon. *Ann. Mo. Bot. Garden* **2019**, *104*, 455–477. [CrossRef]
50. Van Valen, L. Ecological Species, Multispecies, and Oaks. *Taxon* **1976**, *25*, 233–239. [CrossRef]
51. Burger, W.C. The Species Concept in *Quercus*. *Taxon* **1975**, *24*, 45–50. [CrossRef]
52. Hardin, J.W. Hybridization and Introgression in *Quercus Alba*. *J. Arnold Arbor.* **1975**, *56*, 336–363.
53. Kremer, A.; Hipp, A.L. Oaks: An Evolutionary Success Story. *New Phytol.* **2020**, *226*, 987–1011. [CrossRef] [PubMed]
54. Fitzek, E.; Delcamp, A.; Guichoux, E.; Hahn, M.; Lobdell, M.; Hipp, A.L. A Nuclear DNA Barcode for Eastern North American Oaks and Application to a Study of Hybridization in an Arboretum Setting. *Ecol. Evol.* **2018**, *8*, 5837–5851. [CrossRef] [PubMed]
55. Guichoux, E.; Lagache, L.; Wagner, S.; Léger, P.; Petit, R.J. Two Highly Validated Multiplexes (12-Plex and 8-Plex) for Species Delimitation and Parentage Analysis in Oaks (*Quercus* Spp.). *Mol. Ecol. Resour.* **2011**, *11*, 578–585. [CrossRef]
56. Rieseberg, L.H.; Wood, T.E.; Baack, E.J. The Nature of Plant Species. *Nature* **2006**, *440*, 524–527. [CrossRef]
57. Moran, E.V.; Willis, J.; Clark, J.S. Genetic Evidence for Hybridization in Red Oaks (*Quercus* Sect. *Lobatae*, *Fagaceae*). *Am. J. Bot.* **2012**, *99*, 92–100. [CrossRef]
58. Aldrich, P.R.; Parker, G.R.; Michler, C.H.; Romero-Severson, J. Whole-Tree Silvic Identifications and the Microsatellite Genetic Structure of a Red Oak Species Complex in an Indiana Old-Growth Forest. *Can. J. For. Res.* **2003**, *33*, 2228–2237. [CrossRef]
59. Owusu, S.A.; Sullivan, A.R.; Weber, J.A.; Hipp, A.L.; Gailing, O. Taxonomic Relationships and Gene Flow in Four North American Quercus Species (Quercus Section Lobatae). *Syst. Bot.* **2015**, *40*, 510–521. [CrossRef]
60. Scotti-Saintagne, C.; Mariette, S.; Porth, I.; Goicoechea, P.G.; Barreneche, T.; Bodenes, C.; Burg, K.; Kremer, A. Genome Scanning for Interspecific Differentiation between Two Closely Related Oak Species [*Quercus Robur* L. and *Q. Petraea* (Matt.) Liebl.]. *Genetics* **2004**, *168*, 1615–1626. [CrossRef] [PubMed]
61. Garner, M.; Pham, K.K.; Whittemore, A.T.; Cavender-Bares, J.; Gugger, P.F.; Manos, P.S.; Pearse, I.S.; Hipp, A.L. From Manitoba to Texas: A Study of the Population Genetic Structure of Bur Oak (*Quercus Macrocarpa*). *Int. Oaks J. Int. Oak Soc.* **2019**, *30*, 131–138.
62. Hipp, A.L.; Eaton, D.A.R.; Cavender-Bares, J.; Fitzek, E.; Nipper, R.; Manos, P.S. A Framework Phylogeny of the American Oak Clade Based on Sequenced RAD Data. *PLoS ONE* **2014**, *9*, e93975. [CrossRef] [PubMed]
63. Manos, P.S. *Cladistic Analyses of Molecular Variation of "Higher" Hamamelididae and Fagaceae, and Systematics of Quercus Section Protobalanus*; Cornell University: Ithaca, NY, USA, 1993.
64. Manos, P.S. Foliar Trichome Variation in *Quercus* section Protobalanus (Fagaceae). *Sida Contrib. Bot.* **1993**, *15*, 391–403.
65. Ashley, M.V.; Backs, J.R.; Kindsvater, L.; Abraham, S.T. Genetic Variation and Structure in an Endemic Island Oak, Quercus Tomentella, and Mainland Canyon Oak, Quercus Chrysolepis. *Int. J. Plant Sci.* **2018**, *179*, 151–161. [CrossRef]
66. May, M.R.; Provance, M.C.; Sanders, A.C.; Ellstrand, N.C.; Ross-Ibarra, J. A Pleistocene Clone of Palmer's Oak Persisting in Southern California. *PLoS ONE* **2009**, *4*, e8346. [CrossRef]
67. Ortego, J.; Gugger, P.F.; Sork, V.L. Genomic Data Reveal Cryptic Lineage Diversification and Introgression in Californian Golden Cup Oaks (section *Protobalanus*). *New Phytol.* **2018**, *218*, 804–818. [CrossRef]

68. Zumwalde, B.A.; McCauley, R.A.; Fullinwider, I.J.; Duckett, D.; Spence, E.; Hoban, S. Genetic, Morphological, and Environmental Differentiation of an Arid-Adapted Oak with a Disjunct Distribution. *For. Trees Livelihoods* **2021**, *12*, 465. [CrossRef]
69. Tucker, J.M. Studies in the *Quercus undulata* Complex. IV. The Contribution of *Quercus havardii*. *Am. J. Bot.* **1970**, *57*, 71–84. [CrossRef]
70. Bouchal, J.; Zetter, R.; Grímsson, F.; Denk, T. Evolutionary Trends and Ecological Differentiation in Early Cenozoic Fagaceae of Western North America. *Am. J. Bot.* **2014**, *101*, 1332–1349. [CrossRef] [PubMed]
71. Ortego, J.; Gugger, P.F.; Sork, V.L. Climatically Stable Landscapes Predict Patterns of Genetic Structure and Admixture in the Californian Canyon Live Oak. *J. Biogeogr.* **2015**, *42*, 328–338. [CrossRef]
72. Trelease, W. The American Oaks. *Mem. Natl. Acad. Sci.* **1924**, *20*, 1–255.
73. Camus, A.A. *Monographie Du Genre Quercus. Tome I. Genre Quercus. Sous-Genre Cyclobalanopsis et Sous-Genre Euquercus (Section Cerris et Mesobalanu)*; Encyclopédie économique de sylviculture VI.; Editions Paul Lechevalier: Paris, France, 1936.
74. Camus, A.A. *Monographie Du Genre Quercus. Tome II. Genre Quercus. Sous-Genre Euquercus (sections Lepidobalanus et Macrobalanus)*; Encyclopédie économique de sylviculture VII.; Editions Paul Lechevalier: Paris, France, 1938.
75. Camus, A.A. *Monographie Du Genre Quercus. Tome III. Genre Quercus. Sous-Genre Euquercus (sections Protobalanus et Erythrobalanus)*; Encyclopédie économique de sylviculture VIII.; Editions Paul Lechevalier: Paris, France, 1952.
76. Denk, T.; Grimm, G.W. The Oaks of Western Eurasia: Traditional Classifications and Evidence from Two Nuclear Markers. *Taxon* **2010**, *59*, 351–366. [CrossRef]
77. Fick, S.E.; Hijmans, R.J. WorldClim 2: New 1-Km Spatial Resolution Climate Surfaces for Global Land Areas. *Int. J. Climatol.* **2017**, *37*, 4302–4315. [CrossRef]
78. Brown, R. XXX.—Descriptions of Some New or Little-Known Species of Oaks from North-West America. *Ann. Mag. Nat. Hist.* **1871**, *7*, 249–256. [CrossRef]
79. Hipp, A.L. Pharaoh's Dance: The Oak Genomic Mosaic. *Int. Oaks J. Int. Oak Soc.* **2019**, *30*, 53–62.
80. Nixon, K.C. A Biosystematic Study of Quercus Series Virentes (the Live Oaks) with Phylogenetic Analyses of Fagales, Fagaceae and Quercus. Ph.D. Dissertation, University of Texas, Austin, TX, USA, 1985.
81. Cavender-Bares, J.; Pahlich, A. Molecular, Morphological, and Ecological Niche Differentiation of Sympatric Sister Oak Species, *Quercus Virginiana* and *Q. Geminata* (Fagaceae). *Am. J. Bot.* **2009**, *96*, 1690–1702. [CrossRef] [PubMed]
82. Muller, C.H. Relictual Origins of Insular Endemics in Quercus. In Proceedings of the Symposium on the Biology of the Channel Islands; Philbrick, R.N., Ed.; Santa Barbara Botanic Garden: Santa Barbara, CA, USA, 1967; pp. 73–77.
83. Valencia-Avalos, S. Diversidad Del Género *Quercus* (Fagaceae) En México. *Boletín Soc. Botánica México* **2004**, *75*, 33–53.
84. Nixon, K.C. Global and neotropical distribution and diversity of oak (genus Quercus) and oak forests. In *Ecology And Conservation of Neotropical Montane Oak Forests*; Springer: Berlin/Heidelberg, Germany, 2006; pp. 3–13.
85. Avella Muñoz, A.; Rangel Churio, J.O. Oak forests types of *Quercus humboldtii* in the Guantiva-La Rusia-Iguaque Corridor (Santandr-Boyacá, Colombia). *Colomb. For.* **2014**, *17*, 100–116. [CrossRef]
86. Fernández-M., J.F.; Sork, V.L. Genetic Variation in Fragmented Forest Stands of the Andean Oak *Quercus humboldtii* Bonpl. (Fagaceae). *Biotropica* **2007**, *39*, 72–78. [CrossRef]
87. Nixon, K.C. Infrageneric Classification of *Quercus* (Fagaceae) and Typification of Sectional Names. *Ann. Sci. For.* **1993**, *50*, 25–34. [CrossRef]
88. Jensen, R.J. Quercus Linnaeus sect. *Lobatae* Loudon. In *Flora of North America, North of Mexico*; Flora of North America Editorial Committee, Ed.; Oxford University Press: New York, NY, USA, 1997; Volume 3, pp. 447–468.
89. Griswold, E.; Still, S.; McNeil-Marshall, A. Scouting and Collecting Rare Oaks in the Trans-Pecos for Ex-Situ Conservation, 2016. *Int. Oaks J. Int. Oak Soc.* **2018**, *29*, 125–142.
90. Muller, C.H. The oaks of Texas. In *Contributions from Texas Research Foundations*; Lundell, C.L., Ed.; Texas Research Foundation: Renner, TX, USA, 1951; pp. 40–41.
91. Torres-Miranda, A.; Luna-Vega, I.; Oyama, K. Conservation Biogeography of Red Oaks (*Quercus*, Section *Lobatae*) in Mexico and Central America. *Am. J. Bot.* **2011**, *98*, 290–305. [CrossRef] [PubMed]
92. Torres-Miranda, A.; Luna-Vega, I.; Oyama, K. New Approaches to the Biogeography and Areas of Endemism of Red Oaks (*Quercus* L., Section *Lobatae*). *Syst. Biol.* **2013**, *62*, 555–573. [CrossRef]
93. Rodríguez-Correa, H.; Oyama, K. How Are Oaks Distributed in the Neotropics? A Perspective from Species Turnover, Areas of Endemism, and Climatic Niches. *Int. J. Plant Sci.* **2015**, *176*, 222–231. [CrossRef]
94. Keuter, A.; Manos, P.S. Agrifoliae: The California Red Oaks. *Int. Oaks J. Int. Oak Soc.* **2019**, *30*, 191–202.
95. Dodd, R.S.; Afzal-Rafii, Z. Selection and Dispersal in a Multispecies Oak Hybrid Zone. *Evolution* **2004**, *58*, 261–269. [CrossRef] [PubMed]
96. Jensen, R.J. Identifying Oaks: The Hybrid Problem. *J. Int. Oak Soc.* **1995**, *6*, 47–54.
97. Cottam, W.P.; Tucker, J.M.; Santamour, F.S. *Oak Hybridization at the University of Utah*; State Arboretum of Utah: Salt Lake City, UT, USA, 1982.
98. Axelrod, D.I. Biogeography of Oaks in the Arcto-Tertiary Province. *Ann. Mo. Bot. Gard.* **1983**, *70*, 629–657. [CrossRef]
99. Mensing, S. The Paleohistory of California Oaks. In *Proceedings of the Seventh California Oak Symposium: Managing Oak Woodlands in a Dynamic World*; USDA Forest Service General Technical Reports PSW-GTR-251; Pacific Southwest Research Station: Berkeley, CA, USA, 2014; pp. 35–48.

100. Mensing, S. The History of Oak Woodlands in California, Part I: The Paleoecologic Record. *Calif. Geogr.* **2005**, *45*, 1–38.
101. Jensen, R.J. A Preliminary Numerical Analysis of the Red Oak Complex in Michigan and Wisconsin USA. *Taxon* **1977**, *26*, 399–407. [CrossRef]
102. Manos, P.S.; Fairbrothers, D.E. Allozyme Variation in Populations of Six Northeastern American Red Oaks (Fagaceae: *Quercus* Subg. Erythrobalanus). *Syst. Bot.* **1987**, *12*, 265–373. [CrossRef]
103. Guttman, S.I.; Weigt, L.A. Electrophoretic Evidence of Relationships among *Quercus* (oaks) of Eastern North America. *Can. J. Bot.* **1989**, *67*, 339–351. [CrossRef]
104. Hess, W.J.; Stoynoff, N.A. Taxonomic Status of *Quercus acerifolia* (Fagaceae) and a Morphological Comparison of Four Members of the *Quercus shumardii* Complex. *Syst. Bot.* **1998**, *23*, 89–100. [CrossRef]
105. Stoynoff, N.; Hess, W.J. A New Status for *Quercus shumardii* varacerifolia (Fagaceae). *SIDA Contrib. Bot.* **1990**, *14*, 267–271.
106. McDougal, K.M.; Parks, C.R. Elevational Variation in Foliar Flavonoids of *Quercus rubra* L. (Fagaceae). *Am. J. Bot.* **1984**, *71*, 301–308. [CrossRef]
107. Weakley, A.S. Flora of the Southeastern United States. University of North Carolina Herbarium (NCU), North Carolina Botanical Garden: Chapel Hill, NC, USA, 2020.
108. Trelease, W. The Jack Oak (Quercus ellipsoidalis). *Trans. Ill. State Acad. Sci.* **1919**, *12*, 108–118.
109. Jensen, R.J. Lectotypification of *Quercus ellipsoidalis* E.J. Hill. *Taxon* **1979**, *29*, 154–155. [CrossRef]
110. Maycock, P.F.; Daniel, D.R.; Gregory, R.; Reznicek, A.A. Hill's Oak (*Quercus ellipsoidalis*) in Canada. *Can. Field Nat.* **1980**, *94*, 277–285.
111. Ball, P.W. Hills Oak (*Quercus ellipsoidalis*) in Southern Ontario, Canada. *Can. Field Nat.* **1981**, *95*, 281–286.
112. Overlease, W.R. Genetic Relationships between Three Species of Oaks as Determined by Common Garden Studies with Populations from Michigan, Indiana and Wisconsin. *J. Pa. Acad. Sci.* **1991**, *65*, 71–74.
113. Jensen, R.J.; Depiero, R.; Smith, B.K. Vegetative Characters, Population Variation, and the Hybrid Origin of *Quercus ellipsoidalis*. *Am. Midl. Nat.* **1984**, *111*, 364–370. [CrossRef]
114. Wadmond, S.C. The Quercus Ellipsoidalis–Quercus Coccinea Complex. *Trans. Wisc. Acad. Sci. Arts Lett.* **1933**, *28*, 197–203.
115. Overlease, W.R. A Study of the Relationship between Scarlet Oak (*Quercus coccinea* Muenchh.) and Hill Oak (*Quercus ellipsoidalis* E.J. Hill) in Michigan and Nearby States. *J. Pa. Acad. Sci.* **1977**, *51*, 47–50.
116. Shepard, D.A. The Legitimacy of Quercus ellipsoidalis Based on a Populational Study of Quercus Coccinea in Illinois. Western Illinois University: Macomb, IL, USA, 1993.
117. Shepard, D.A. A Review of the Taxonomic Status of *Quercus ellipsoidalis* and *Quercus coccinea* in the Eastern United States. *Int. Oak J.* **2009**, *20*, 65–84.
118. Hipp, A.L.; Weber, J.A.; Srivastava, A. Who Am I This Time? The Affinities and Misbehaviors of Hill's Oak (*Quercus ellipsoidalis*). *Int. Oak J.* **2010**, *21*, 27–36.
119. Zhang, R.; Hipp, A.L.; Gailing, O. Sharing of Chloroplast Haplotypes among Red Oak Species Suggests Interspecific Gene Flow between Neighboring Populations. *Botany* **2015**, *93*, 691–700. [CrossRef]
120. Hokanson, S.C.; Isebrands, J.G.; Jensen, R.J.; Hancock, J.F. Isozyme Variation in Oaks of the Apostle Islands in Wisconsin: Genetic Structure and Levels of Inbreeding in *Quercus rubra* and *Q. ellipsoidalis* (Fagaceae). *Am. J. Bot.* **1993**, *80*, 1349–1357. [CrossRef]
121. Sullivan, A.R.; Owusu, S.A.; Weber, J.A.; Hipp, A.L.; Gailing, O. Hybridization and Divergence in Multi-Species Oak (*Quercus*) Communities. *Bot. J. Linn. Soc.* **2016**, *181*, 99–114. [CrossRef]
122. Khodwekar, S.; Gailing, O. Evidence for Environment-Dependent Introgression of Adaptive Genes between Two Red Oak Species with Different Drought Adaptations. *Am. J. Bot.* **2017**, *104*, 1088–1098. [CrossRef] [PubMed]
123. Lind-Riehl, J.F.; Sullivan, A.R.; Gailing, O. Evidence for Selection on a CONSTANS-like Gene between Two Red Oak Species. *Ann. Bot.* **2014**, *113*, 967–975. [CrossRef]
124. Noss, R.F.; Platt, W.J.; Sorrie, B.A.; Weakley, A.S.; Means, D.B.; Costanza, J.; Peet, R.K. How Global Biodiversity Hotspots May Go Unrecognized: Lessons from the North American Coastal Plain. *Divers. Distrib.* **2015**, *21*, 236–244. [CrossRef]
125. Hunt, D.M. *A Systematic Review of Quercus Series Laurifoliae, Marilandicae and Nigrae*; University of Georgia: Athens, GA, USA, 1990.
126. Ortego, J.; Gugger, P.F.; Riordan, E.C.; Sork, V.L. Influence of Climatic Niche Suitability and Geographical Overlap on Hybridization Patterns among Southern Californian Oaks. *J. Biogeogr.* **2014**, *41*, 1895–1908. [CrossRef]
127. Kim, B.Y.; Wei, X.; Fitz-Gibbon, S.; Lohmueller, K.E.; Ortego, J.; Gugger, P.F.; Sork, V.L. RADseq Data Reveal Ancient, but Not Pervasive, Introgression between Californian Tree and Scrub Oak Species (*Quercus* Sect. *Quercus*: Fagaceae). *Mol. Ecol.* **2018**, *27*, 4556–4571. [CrossRef] [PubMed]
128. Tucker, J.M. Studies in the *Quercus undulata* Complex. I. A Preliminary Statement. *Am. J. Bot.* **1961**, *48*, 202. [CrossRef]
129. Howard, D.J.; Preszler, R.W.; Williams, J.; Fenchel, S.; Boecklen, W.J. How Discrete Are Oak Species? Insights from a Hybrid Zone between *Quercus grisea* and *Quercus gambelii*. *Evolution* **1997**, *51*, 747–755. [CrossRef]
130. Maze, J. Past Hybridization between *Quercus macrocarpa* and *Quercus gambelii*. *Brittonia* **1968**, *20*, 321–333. [CrossRef]
131. Leroy, T.; Roux, C.; Villate, L.; Bodénès, C.; Romiguier, J.; Paiva, J.A.P.; Dossat, C.; Aury, J.-M.; Plomion, C.; Kremer, A. Extensive Recent Secondary Contacts between Four European White Oak Species. *New Phytol.* **2017**, *214*, 865–878. [CrossRef]
132. Nixon, K.C.; Muller, C.H. Quercus Linnaeus sect. Quercus. In *Flora of North America*; Flora of North America Editorial Committee, Ed.; Oxford University Press: New York, NY, USA, 1997; Volume 3, pp. 471–506.

133. Tucker, J.M. Studies in the *Quercus undulata* Complex. III. The Contribution of *Q. arizonica. Am. J. Bot.* **1963**, *50*, 699–708. [CrossRef]
134. Swenson, N.G.; Fair, J.M.; Heikoop, J. Water Stress and Hybridization between *Quercus gambelii* and *Quercus grisea. West. N. Am. Nat.* **2008**, *68*, 498–507. [CrossRef]
135. Nixon, K.C.; Steele, K.P. A New Species of *Quecus* (Fagaceae) from Southern California. *Madroño* **1981**, *28*, 210–219.
136. Rundel, P.W.; Arroyo, M.T.K.; Cowling, R.M.; Keeley, J.E.; Lamont, B.B.; Vargas, P. Mediterranean Biomes: Evolution of Their Vegetation, Floras, and Climate. *Annu. Rev. Ecol. Evol. Syst.* **2016**, *47*, 383–407. [CrossRef]
137. Burge, D.O.; Parker, V.T.; Mulligan, M.; Sork, V.L. Influence of a Climatic Gradient on Genetic Exchange between Two Oak Species. *Am. J. Bot.* **2019**, *106*, 864–878. [CrossRef]
138. Hardin, J.W. Patterns of Variation in Foliar Trichomes of Eastern North American *Quercus. Am. J. Bot.* **1979**, *66*, 576–585. [CrossRef]
139. Thomson, P.M.; Mohlenbrock, R.H. Foliar Trichomes of *Quercus* Subgenus *Quercus* in the Eastern United States. *J. Arnold Arbor.* **1979**, *60*, 350–366.
140. Duncan, W.H. *Quercus oglethorpensis*—Range Extensions and Phylogenetic Relationships. *Lloydia* **1950**, *13*, 243–248.
141. Nixon, K.C.; Muller, C.H. The Taxonomic Resurrection of *Quercus laceyi* Small (Fagaceae). *SIDA Contrib. Bot.* **1992**, *15*, 57–69.
142. Axelrod, D.I. Evolution of the Madro-Tertiary Geoflora. *Bot. Rev.* **1958**, *24*, 433–509. [CrossRef]
143. Van Devender, T.R.; Spaulding, W.G. Development of Vegetation and Climate in the Southwestern United States. *Science* **1979**, *204*, 701–710. [CrossRef] [PubMed]
144. Backs, J.R. Genetic Analysis of a Rare Isolated Species: A Tough Little West Texas Oak, *Quercus hinckleyi* C.H. Mull. *J. Torrey Bot.* **2015**, *142*, 302–313. [CrossRef]

Article

Genome-Wide Variation in DNA Methylation Predicts Variation in Leaf Traits in an Ecosystem-Foundational Oak Species

Luke Browne [1,2], Brandon MacDonald [1], Sorel Fitz-Gibbon [1], Jessica W. Wright [3] and Victoria L. Sork [1,4,*]

1 Department of Ecology and Evolutionary Biology, Charles E. Young Drive, University of California, Los Angeles, CA 90095-1438, USA; luke.browne@yale.edu (L.B.); brandonwsmacdonald@gmail.com (B.M.); sorel@ucla.edu (S.F.-G.)
2 UCLA La Kretz Center for California Conservation Science, Institute of the Environment and Sustainability, University of California, Los Angeles, CA 90095-1496, USA
3 USDA-Forest Service Pacific Southwest Research Station, 1731 Research Park Dr., Davis, CA 95618, USA; jessica.w.wright@usda.gov
4 Institute of the Environment and Sustainability, University of California, Los Angeles, CA 90095-1496, USA
* Correspondence: vlsork@ucla.edu

Abstract: Epigenetic modifications such as DNA methylation are a potential mechanism for trees to respond to changing environments. However, it remains controversial the extent to which DNA methylation impacts ecologically important traits that influence fitness. In this study, we used reduced-representation bisulfite sequencing to associate genomic and epigenomic variation with seven phenotypic traits related to growth, leaf function, and disease susceptibility in 160 valley oak (*Quercus lobata*) saplings planted across two common gardens in California. We found that DNA methylation was associated with a significant fraction of phenotypic variance in plant height, leaf lobedness, powdery mildew infection, and trichome density. Two of the seven traits were significantly associated with DNA methylation in the CG context, three traits were significantly associated with CHG methylation, and two traits were significantly associated with CHH methylation. Notably, controlling for genomic variation in SNPs generally reduced the amount of trait variation explained by DNA methylation. Our results suggest that DNA methylation may serve as a useful biomarker to predict phenotypic variation in trees, though it remains unclear the degree to which DNA methylation is a causal mechanism driving phenotypic variation in forest tree species.

Keywords: phenotype; DNA methylation; *Quercus*; single nucleotide polymorphisms; functional traits; leaf traits; bisulfite sequencing

Citation: Browne, L.; MacDonald, B.; Fitz-Gibbon, S.; Wright, J.W.; Sork, V.L. Genome-Wide Variation in DNA Methylation Predicts Variation in Leaf Traits in an Ecosystem-Foundational Oak Species. *Forests* **2021**, *12*, 569. https://doi.org/10.3390/f12050569

Academic Editors: Mary Ashley and Janet R. Backs

Received: 26 March 2021
Accepted: 29 April 2021
Published: 1 May 2021

Publisher's Note: MDPI stays neutral with regard to jurisdictional claims in published maps and institutional affiliations.

1. Introduction

Due to rapidly changing environmental conditions caused by drivers of global change, understanding the potential role of epigenetics in plant response to the environment has become a topic of interest for ecologists and evolutionary biologists [1–5]. DNA methylation, where cytosine bases are methylated by distinct molecular pathways in either the CG, CHG, or CHH (H = C, A, or T) sequence contexts, is a prominent epigenetic modification [4,6,7] that may play a role in adaptive phenotypic responses to the environment [8–11]. In addition, phenotypic traits mediated by DNA methylation can be heritable [10,12–16], providing the raw material for evolution to occur and a mechanism for induced phenotypic responses to the environment to persist across generations. Several studies have shown DNA methylation to be associated with phenotypic traits, such as drought tolerance in *Arabidopsis* [10], flowering time in flax [17], adaption to temperature in *Arabidopsis* [18], and various traits in other herbaceous species [10,17,19,20]. However, studies resolving the associations between DNA methylation and phenotype are less common in tree species despite the advantages epigenetic modifications may bring for plants with long lifespans to be able to respond phenotypically to rapid climate change [21].

Forest tree species are among the most economically and ecologically important plant taxa globally, and understanding their epigenetic response to changing conditions is critical [22–24]. In particular, oaks (*Quercus* spp.) are foundational members of many ecosystems, having higher biomass and species diversity than any other woody genus in the United States and Mexico [25]. Previous studies of DNA methylation in oaks have explored natural variation across populations in *Quercus lobata* [26,27], drought stress in *Quercus ilex* [28], temperature stress in *Quercus suber* [29], cork quality in *Quercus suber* [30], and seedling growth and gene expression in *Quercus lobata* [31]. These studies are complemented by research in other woody tree species, such as *Picea abies* [32], *Populus trichocarpa* [33], *Eucalpytus* [34], *Pistacia lentiscus* [35], and *Ilex aquifolium* [36] ranging in focus from the effects of temperature on embryogenesis, drought stress, epigenetic differences among clonal genotypes, and associations between epigenetic variation and leaf phenotype. Continuing to study the role of epigenetic modifications in woody species is critical to assessing the extent to which DNA methylation shapes plant phenotype [21].

If DNA methylation is providing a mechanism for plant response to the environment, there must be a causal link between variation in DNA methylation and plant phenotype. In some cases, DNA methylation is known to play a causal role in regulating gene expression and variation in phenotypic traits [37–39]. However, the general connection between DNA methylation and phenotype remains unresolved [40]. For example, in one *Arabidopsis* study, gene expression did not change consistently with the loss of gene body methylation (gbM [41]), which is consistent with other evidence that gbM shows no functional relationship with gene expression. However, analysis of multiple accessions of *Arabidopsis* indicates that gbM is a heritable trait "consistent with the hypothesis that gbM is the result of a passive process related to the variation in the robustness of the maintenance of heterochromatin", which may lead to the evolution of epialleles that shape phenotypes [16]. In tomato, experimental studies showed that grafting root stock with epigenetic variation associated with msh1 affected growth that was heritable over five generations [42]. An important first step for natural tree populations would be to establish an association (not necessarily causal) between phenotypic variation at ecologically important plant traits and variation in DNA methylation.

Variation in DNA methylation may or may not be independent from underlying genetic variation in DNA sequence, which further complicates efforts to resolve the role of DNA methylation on phenotype. Statistically, it is possible to account for this connection by controlling for DNA sequence variation, often in the form of SNPs (single nucleotide polymorphisms), in any association analyses between DNA methylation and phenotype. Genomic relationship matrices are commonly used to estimate the proportion of phenotypic variance in a trait explained by all measured SNP or DNA methylation markers [43,44]. This approach works especially well for complex, polygenic traits that are controlled by many sites across the genome, although SNPs have limitations. For example, transposable element insertion polymorphisms (TIPS) can also shape phenotypic diversity and these many not be associated with SNPs [45]. Nonetheless, it is instructive to analyze the genomic and epigenomic basis of complex traits through simultaneous estimation of phenotypic variance from relationship matrices generated from both SNPs and DNA methylation markers [46,47]. In this way, it is possible estimate the association between DNA methylation and phenotypic traits while controlling for potential underlying genetic variation.

In this study, we tested the association between DNA methylation and phenotypic traits in a widely distributed, endemic California oak, *Quercus lobata* (valley oak) as a first step in finding candidate phenotypes potentially influenced by epigenetic processes. Previous studies of valley oak have revealed strong associations between climate gradients and variation in DNA methylation across the genome [26,27] and reduced seedling growth and altered gene expression when DNA methylation is inhibited experimentally [31]. Moreover, oaks have an unusual pattern of DNA methylation where subcontext-specific patterns of DNA methylation associated with transposable elements reveal that the genome has broadly distributed heterochromatin in intergenic regions [48], which is a pattern more similar to grasses than to other angiosperms [49]. Using data from 160 valley oak saplings planted across two common gardens in California, we addressed the following questions: Is there an association between DNA methylation and plant growth and leaf traits? How does this relationship vary by sequence context? If we control for the underlying genetic variation across individuals, do associations between DNA methylation and phenotypic traits remain? In addition, we perform a series of data simulations to test the power and precision of our analysis approach in an effort to guide future studies.

2. Materials and Methods

2.1. Study Species

Quercus lobata Née (valley oak) is an ecosystem-foundational oak species endemic to California that occurs along the foothills of the Sierra Nevada and Coastal mountain ranges (Figure 1). As a dominant species in woodlands and savannas, valley oak has high ecological importance [50], in addition to its high cultural value to Native Americans [51]. In general, valley oak populations have high local genetic diversity, low genetic structure, and significant associations between genotypic and climatic gradients across its range [52,53]. Previous studies of valley oak suggest that DNA methylation may be involved in local adaptation to climate, especially in the CG context, which shows high differentiation among populations and strong associations with temperature gradients [26,27].

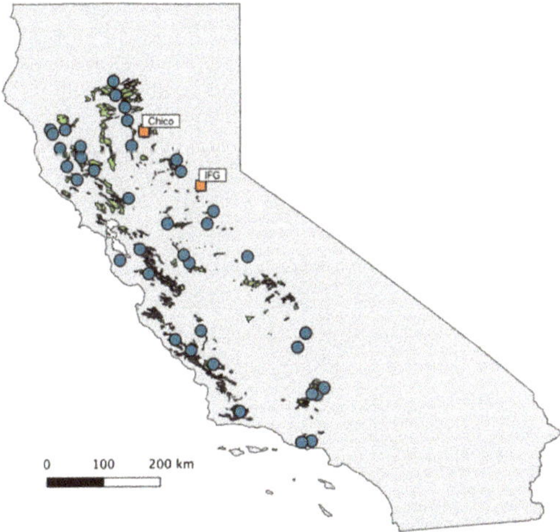

Figure 1. Map of provenances (blue circles) individuals sampled in this study across the range of valley oak in California (green polygons). Orange rectangles show location of the two common gardens in Chico and Placerville (IFG), California.

2.2. Valley Oak Provenance Trial

All phenotypic and sequencing data used in this study come from the valley oak provenance trial, and the full details on the design and establishment of the valley oak provenance trial are available in [54]. In 2012, >10,000 open-pollinated acorns were collected from 674 adult valley oaks in 95 localities distributed across the species range. These acorns were germinated in a greenhouse and in 2014, 6945 seedlings were planted at two sites in Placerville (IFG) and Chico, California (Figure 1). Seedlings were planted with 2.25 m spacing intervals in a blocked design, such that each family (i.e., half-sib progeny collected as acorns from a single adult) had at least one individual in each block at each site. Both sites were irrigated and weeded to increase the probability of seedling establishment, as is commonly done in provenance trials.

2.3. Phenotypic Measurements

In 2016, when saplings were approximately four years old, we measured a set of seven phenotypic traits related to performance and leaf function: height of the tallest stem (cm), leaf area (cm^2), leaf lobedness, specific leaf area (SLA, cm^2 g^{-1}), leaf thickness (mm), trichome density (categorical, see below), and powdery mildew infection intensity (categorical, see below) [55]. Height of the tallest stem in valley oak seedlings is strongly correlated with other metrics of size and growth, such as diameter at breast height (dbh), and likely plays a key role in early stage survival for avoiding herbivory and surviving abiotic stress [56]. For leaf traits, we sampled mature leaves that were ~10 cm from the branch tip from each side of the tree. We used one leaf sample per individual for estimating trichome density, two samples for leaf thickness and 4–5 leaves per individual for leaf area, specific leaf area, and leaf lobedness. Pairwise correlations among phenotypic traits were all $R < 0.46$ (Table 1).

Table 1. Pairwise correlations between phenotypic traits measured on valley oak individuals in this study. The bottom diagonal shows the estimated Pearson's correlation coefficient (R) between pairs of traits.

Trait	Height	Leaf Area	Leaf Lobedness	Specific Leaf Area	Leaf Thickness	Trichome Density	Powdery Mildew
Height	1.00						
Leaf area	0.46	1.00					
Leaf lobedness	0.25	0.34	1.00				
Specific leaf area	−0.14	0.04	0.01	1.00			
Leaf thickness	−0.15	0.18	−0.10	−0.19	1.00		
Trichome density	0.29	−0.06	−0.19	−0.19	−0.05	1.00	
Powdery mildew	0.44	0.17	−0.05	−0.06	−0.11	0.16	1.00

We measured leaf area by scanning fully hydrated leaves and estimating area using the ImageJ software [57]. To estimate leaf lobedness, we estimated leaf perimeter using ImageJ and then divided leaf perimeter^2 by leaf area. Leaf lobedness in oaks is negatively associated with hydraulic resistance and may be a mechanism for optimizing water balance under dry conditions [58]. We then pressed and dried the leaves to calculate SLA (leaf area divided by dried mass) [59]. Higher SLA has been shown to be related to higher leaf nitrogen concentration, higher potential relative growth rates, and lower leaf longevity and lower investment in secondary compounds [59]. We measured leaf thickness at the midpoint between the perimeter and midrib, midway between the base and the tip using digital calipers, making sure to avoid secondary veins, as recommended by [59]. Higher leaf thickness is related to longer leaf longevity, higher toughness, higher photosynthetic capacity per unit leaf area, and leaf nitrogen per unit area, which may be advantageous when under high levels of moisture availability [59,60]. We estimated trichome density by viewing the abaxial side of the leaf under a microscope at 1500× magnification following [61] and [62] and categorizing trichome density into 6 categories: <15, 15–24, 25–34, 35–44, 45–54, and ≥55 trichomes in the field of view. Higher trichome densities reduce water loss by decreasing the rate of transpiration and may be advantageous in dry conditions [63]. We categorized powdery mildew infection based on the proportion of leaves on the entire sapling displaying signs of infection: no signs of infection, low infection (<33% of leaves infected), medium infection (33–66% of leaves infected), high infection (>66% of leaves displaying sign of infection).

We analyzed the relationship between SNP, methylation, and phenotypic variation using both raw phenotypic trait values and phenotypic trait values adjusted to control for local environmental effects of each common garden. Raw phenotypic values include the potential contribution of environmentally induced DNA methylation on phenotypic traits, while the adjusted phenotypic values include the contribution of DNA methylation to a phenotypic trait after controlling for local environmental effects. To accomplish this, we fit a separate linear regression for each phenotypic trait using the phenotypic trait as the response variable and common garden location and block nested within common garden location as random effects using the 'lmer' function in the 'lme4' R package [64]. We then took the residuals of each model as the adjusted phenotypic values.

2.4. Bisulfite Sequencing

We chose 160 focal saplings for bisulfite sequencing, where we simultaneously identified genomic SNPs and variation in DNA methylation across the genome using the same DNA sample (see below). The saplings came from a total of 40 maternal families, sampling 2 progeny from each family in each planting site. We followed the same bisulfite sequencing protocol described elsewhere [27]. Briefly, we extracted total genomic DNA from leaf tissue using a prewash protocol [65] and the Qiagen Dneasy plant extraction kit. We modified the protocol of Feng et al. [66] to prepare reduced-representation bisulfite sequencing (RRBS) libraries. We first digested total genomic DNA with *Mspl* (CCGG) that was then end-repaired and adenylated with Klenow fragments (3' to 5' exo-). We ligated Illumina TruSeq adapters to fragments in each library. We performed size selection using AMPureXP beads to target fragments of 200 to 500 bp (including ~120 bp of adapter sequence). We treated libraries with sodium bisulfite (EpiTect, Qiagen) to perform the conversion of unmethylated cytosines to uracil, which then are subsequently read as thymine during sequencing. The prepared libraries were PCR amplified with Illumina primers and then pooled in batches of 12. Pooled libraries were sequenced on multiple lanes on an Illumina HiSeq 2000 v3 in 100-bp single end mode at the core facilities of the Broad Stem Cell Center, UCLA.

2.5. Methylation and SNP Calling

We filtered Illumina reads that failed the Illumina chastity test and then converted from QSEQ to FASTQ format, demultiplexed by sample, and applied the EAMSS quality score correction. We used TrimGalore (https://github.com/FelixKrueger/TrimGalore, accessed on 31 October 2019) in 'rrbs' mode to trim 3–4 bases at the ends of reads after examining mbias plots generated in MethylDackel (https://github.com/dpryan79/MethylDackel, accessed on 31 October 2019), ensuring reads were at least 20 bases with a quality score of 20. We aligned trimmed reads using bwa-meth [67] to v3.0 of the valley oak genome [68] using default settings. We extracted the methylation levels separately for the CG, CHG, and CHH context using MethylDackel, with minimum mapping quality of 40, a minimum depth of coverage of $10\times$, and a maximum depth of coverage of $1000\times$. These methylation levels range from 0 to 1 and estimate the fraction of cells containing DNA methylation at a locus for a given sample. We filtered methylation sites such that each site had <10% missing data and had a standard deviation >0.01, to select for variable sites. We then removed any methylation site that had a high correlation ($R^2 \geq 0.70$) with any other methylation site in the same sequence context within 500 bp until no such pairs remained, leaving a set of uncorrelated methylation sites. To reduce the influence of extreme methylation levels at individual samples in analysis, we winsorized methylation levels at each site using the 'Winsorize' function in the DescTools R package [69]. Winsorizing limits extreme values in a dataset and reduces the effect of spurious outliers by setting all data below the 5th percentile to the 5th percentile and all data above the 95th percentile to the 95th. Finally, we removed two individual valley oaks for having $\geq 25\%$ missing data across all methylation sites. After filtering, we obtained 6175 mCG, 2831 mCHG, and 4350 mCHH variable and high-quality methylation sites across the valley oak genome.

To call SNPs from the RRBS reads [70], we used the BISulfite-seq CUI Toolkit (BISCUIT) (https://github.com/zhou-lab/biscuit, accessed on October 31 2019). We set a minimum base quality of 20 and mapping quality of 20 during SNP calling. We then used the Genome Analysis Tool Kit [71] to filter variants with an allele count AC < 4, allele frequency AF < 0.10, and genotype quality GQ < 20. We further filtered SNPs using vcftools [72] such that all SNPs were biallelic, had < 10% missing data, minor allele frequency > 0.05, and a mean depth of coverage over all individuals $\geq 5\times$ and $\leq 500\times$. To reduce the occurrence of spurious SNP calls from the RRBS reads, we used vcftools to exclude variants that were out of Hardy–Weinberg Equilibrium with $p < 0.0000001$. Then, we used plink [73] to filter for linkage disequilibrium among SNPs, similar to the methylation sites, where a SNP was removed if it was highly correlated ($R^2 \geq 0.70$) with any other SNP in a 500 bp window until no such pairs remain. After filtering, we retained 8450 high quality and variable SNPs.

2.6. Trait Associations

To determine the association between DNA methylation and our set of phenotypic traits, we used the GCTA (Genome-wide Complex Trait Analysis) software [47]. GCTA estimates the proportion of phenotypic variance of a trait explained by genomic data (e.g., SNP or DNA methylation data) using relatedness matrices and genome-based restricted maximum likelihood (GREML) [44]. The use of relatedness matrices to explain variation in complex traits has grown recently in ecology and evolutionary biology [43,46] due to the increasing availability of reduced representation sequencing that is well-suited for these types of analyses. Additionally, GCTA is able to estimate the separate contributions of multiple relatedness matrices to phenotypic variance. In this way, we were able to separate the amount of phenotypic variance explained by DNA methylation after controlling for underlying genetic variation in the sampled SNPs. Importantly though, SNPs outside of the sampled sequences may still play an important role in predicting phenotypic variation, which would not be accounted for due to lack of sampling. The approach does not relate individual SNPs or methylation sites to individual traits, but rather uses variation in SNPs and DNA methylation sampled across the entire genome to predict phenotype.

We calculated relatedness matrices from the SNP and DNA methylation for each sequence context separately (e.g., CG, CHG, CHH) using OSCA (OmicS-data-based Complex trait Analysis) [74]. In all model runs, we included the first 2 principal components of the relatedness matrix or matrices included in the analysis to help control for population stratification (Figure S1). We also included mean depth of coverage to control for differences in sequencing coverage among individuals. Mean depth of coverage across sites was highly correlated across SNP, CG, CHG, and CHH sites (Pearson's $R > 0.98$ for all comparisons), and thus for simplicity we only used depth of coverage at SNP sites as a covariate in the GCTA analysis. We used the expectation-maximization restricted maximum likelihood (EM-REML) algorithm option with no constraint and 10,000 maximum iterations to aid in model convergence.

We included 154 saplings in the final trait analyses (two excluded for missing methylation data and four excluded for missing trait data). With this sample size, we expect high uncertainty and low power to detect trait associations. However, our goal in this study was to determine if there is a detectable association between DNA methylation, SNPs, and phenotypic traits, rather than precisely estimate the amount of variance DNA methylation can explain, which would require much larger sample sizes. To this end, we used a permutation-based approach to test whether the amount of variance explained by SNPs or DNA methylation in each phenotypic trait was higher than would be expected by chance, under the null hypothesis that there was no association between SNPs, DNA methylation, and our phenotypic traits, following the approach of [75]. To accomplish this, for each model run, we randomized phenotypes among individuals $n = 10,000$ times and in each iteration calculated the percent of variance explained by DNA methylation and SNPs in permuted phenotypes. We compared this null distribution of expected percent variance explained in phenotypes with the observed phenotypic variance explained in the observed data. If the observed amount of phenotypic variance was >95% of permuted values, we interpret this as a signal of a significant association between the trait of interest and DNA methylation of SNP variation. If more than one relatedness matrix was included in the model run, we separately estimated the null distribution of phenotypic variance explained for each matrix.

Additionally, we performed a series of data simulations to test the validity and power of our permutation-based approach. We simulated a series of different scenarios that varied the number of individuals sampled and the amount of trait variance explained by either SNP or DNA methylation variation. In each simulation repetition, we first simulated a SNP dataset with 1000 markers all with minor allele frequency >0.05. We then simulated a set of 1000 DNA methylation markers that varied in methylation levels from 0.0–1.0 at each marker. We next simulated independent polygenic additive effect sizes for each SNP and methylation marker that was then used to simulate a phenotypic trait. The effect sizes of SNP and methylation markers were independent from each other as well. We varied the true amount of variance in the phenotypic trait explained by either the SNP or DNA methylation markers from 0.00001 (required to be a non-zero number), 0.10, 0.20, 0.30, 0.40. We also varied the number of individuals being simulated from 50, 150, 250, 500, to 1000 individuals. For each parameter combination, we employed the permutation-based testing approach described above for detecting a significant effect of either SNP or DNA methylation in predicting trait variation. We simulated 100 repetitions of each parameter combination. We combined the results to calculate false positive rates when DNA methylation explained close to 0 trait variance and the statistical power of the permutation-based approach when a non-zero amount of trait variance was explained by DNA methylation across sample sizes.

3. Results

Overall, DNA methylation alone in any context without controlling for genetic variation explained a significant fraction of variance in four of the seven traits we measured (Figure 2). However, the amount of trait variation explained by DNA methylation was generally reduced and often was no longer statistically significant after controlling for genetic variation (Figure 2). The few exceptions to this pattern were CHG and CHH methylation predicting plant height and CHG methylation predicting extent of powdery mildew infection, which both explained a significant fraction of trait variation even after controlling for genetic variation (Figure 2). Traits varied in which DNA methylation contexts were able to explain variation, with some traits associated with CG methylation (e.g., leaf lobedness, trichome density), others with CHG methylation (e.g., height, extent of powdery mildew infection, trichome density), and others with CHH methylation (e.g., height, trichome density, Figure 2). Leaf area, specific leaf area, and leaf thickness did not show significant associations with DNA methylation in any context (Figure 2). SNP variation explained a significant fraction of trait variation for leaf area, leaf lobedness, and trichome density (Figure 2).

Adjusting phenotypic trait values for local environmental effects (i.e., controlling for variation across common garden sites and among blocks within sites) affected estimates of the amount of variation explained primarily for height, trichome density, and extent of powdery mildew infection (Figure 2). In these cases, adjusted phenotypes showed weaker associations with variation in DNA methylation when local environmental effects were accounted for (Figure 2). The amount of trait variation explained by SNPs was not affected by controlling for local environmental effects (Figure 2).

Using simulated data, we found that our analysis approach of permuting phenotypes to simulate the null hypothesis of no association between SNP or DNA methylation data and traits had an acceptable false discovery rate (~5%) for testing the null hypothesis of no association between trait variation and SNP/DNA methylation variation at sample sizes similar to those used in this study (n = 150 individuals, Figure 3). However, the analysis approach had low power to detect associations (e.g., 20% power to detect association if DNA methylation explains 30% of trait variance) at these sample sizes (Figure 3). Furthermore, we found high levels of variation and uncertainty in the estimates of variation explained when sample sizes were <500 individuals (Figure S2).

Trait and DNA methylation / SNP associations
(percent variance explained)

	SNP	mCG	mCG I SNP	mCHG	mCHG I SNP	mCHH	mCHH I SNP
Height	1% (0.22)	3% (0.35)	2% (0.6)	94%* (0.36)	96%* (0.39)	76%* (0.45)	79%* (0.47)
Height (adjusted)	2% (0.22)	6% (0.35)	4% (0.6)	12% (0.39)	16% (0.43)	7% (0.46)	6% (0.49)
Leafarea	47%* (0.25)	48% (0.37)	9% (0.6)	18% (0.39)	7% (0.41)	3% (0.46)	2% (0.47)
Leafarea (adjusted)	59%* (0.25)	58% (0.38)	5% (0.59)	16% (0.39)	3% (0.4)	2% (0.46)	1% (0.46)
Lobedness	60%* (0.25)	87%* (0.37)	53%* (0.58)	14% (0.39)	9% (0.4)	9% (0.47)	6% (0.45)
Lobedness (adjusted)	76%* (0.24)	91%* (0.37)	31% (0.57)	10% (0.39)	4% (0.38)	10% (0.47)	5% (0.44)
Mildew	2% (0.22)	6% (0.35)	4% (0.6)	64%* (0.38)	63%* (0.42)	20% (0.47)	22% (0.49)
Mildew (adjusted)	9% (0.23)	12% (0.36)	7% (0.61)	42% (0.39)	34% (0.43)	4% (0.46)	4% (0.49)
Sla	17% (0.24)	44% (0.37)	38% (0.61)	58% (0.39)	49% (0.42)	24% (0.47)	21% (0.49)
Sla (adjusted)	23% (0.24)	50% (0.37)	39% (0.61)	63% (0.38)	53% (0.42)	26% (0.47)	23% (0.49)
Thickness	7% (0.23)	5% (0.35)	4% (0.6)	7% (0.39)	5% (0.42)	10% (0.47)	6% (0.49)
Thickness (adjusted)	12% (0.23)	5% (0.35)	4% (0.6)	4% (0.39)	3% (0.42)	7% (0.46)	4% (0.49)
Trichome	90%* (0.22)	76%* (0.37)	4% (0.55)	84%* (0.37)	21% (0.38)	85%* (0.44)	17% (0.43)
Trichome (adjusted)	93%* (0.22)	85%* (0.37)	4% (0.55)	69%* (0.38)	5% (0.37)	38% (0.47)	2% (0.41)

Figure 2. Percent of variance explained in seven traits of valley oak by variation in SNPs and each DNA methylation context. 'mCG I SNP', 'mCHG I SNP', and 'mCHH I SNP' indicate the percent variance explained by each methylation context after controlling for genetic variation in the SNPs. Standard errors are indicated in parentheses. Trait values were either the raw phenotypic values or the adjusted values after controlling for site and block effects in the common gardens. Bolded values with asterisks indicate significant associations ($p < 0.05$) from a permutation-based approach. Darker shadings of blue indicate higher values of percent variance explained.

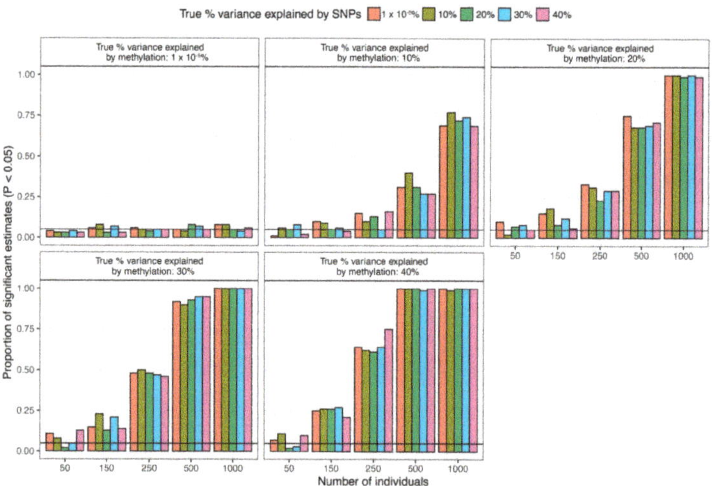

Figure 3. Statistical power increases with sample sizes and is generally not affected by the amount of variance explained by SNPs. Proportion of significant estimates of the relationship between simulated DNA methylation and simulated phenotypic data based on varying sample sizes and varying the true amount of variance explained by methylation and the true amount of variance explained by SNPs. Horizontal line shows where proportion of significant estimates is 0.05.

4. Discussion

Overall, we found significant associations between DNA methylation and phenotype for four of the seven traits measured in this study (plant height, leaf lobedness, disease susceptibility, and trichome density). The phenotypic traits varied in their association with DNA methylation based on sequence context, with traits such as height associated with mCHG and mCHH methylation, while leaf lobedness associated most strongly with mCG sites. In addition, the amount of variation explained by DNA methylation was substantially reduced in some cases after controlling for genetic variation in SNPs, indicating that variation in DNA methylation and genomic background may be confounded in some instances. Resolving the associations between methylation and phenotypic traits is important for understanding the role of epigenetic modifications in plant response to the environment. This study provides one of the few examples of significant associations between DNA methylation and phenotype for a tree species.

The height of 4-year-old valley oak saplings showed strong associations with variation in mCHG and mCHH methylation and no associations with SNPs or mCG methylation. When we adjusted height to control for local environmental effects (i.e., differences within and across common gardens), the significant association between height and mCHG and mCHH no longer remained. In general, plant height and growth are plastic traits largely influenced by environmental conditions experienced by individual plants, though growth in valley oak can be partially explained by SNP variation [76]. The fact that the association between DNA methylation and height was removed after controlling for plastic phenotypic responses suggests mCHG and mCHH methylation exhibit plastic responses to the environment even within common gardens, which correspond to changes in plant height. A previous experimental study where valley oak seedlings that were chemically demethylated suggests a potential mechanistic link between DNA methylation and plant height in valley oaks [31]. When DNA methylation was reduced in all sequence contexts, seedlings showed a ~50% reduction in new growth compared to controls [31]. Furthermore, associations between DNA methylation and plant growth have been demonstrated in other systems, including *Eucalyptus* [34,77,78], suggesting that plant growth may be a phenotypic trait particularly sensitive to variation in DNA methylation. At the same time, plant growth is also a fitness component which may reflect the consequences of traits on plant performance rather than a direct effect on polygenic traits associated with growth.

Leaf lobedness, a trait associated with leaf hydraulic resistance [58], showed strong associations with variation in both SNPs and mCG methylation, but no association with mCHG or mCHH methylation. Leaf lobedness has been shown to vary across natural populations spanning the valley oak range [79] and to be positively associated with photosynthetic rates [80], and thus might be a target of divergent selection. Our results support the idea that leaf lobedness is a heritable trait where variation in SNPs explain a significant amount of variation. In addition, mCG methylation by itself explained variation in leaf lobedness, though this amount was reduced when also controlling for variation in SNPs. Given that mCG methylation itself can be heritable, it seems likely that the association between mCG methylation and leaf lobedness is driven by an underlying correlation between variation in mCG methylation and SNPs. Because most CG methylation is in gene bodies and it has been shown that gene body methylation may not directly influence gene expression, the reason for this association will need more investigation. Indeed, the underlying causal mechanism driving phenotypic variation is likely to be more associated with the genomic background rather than mCG methylation itself. Similarly, trichome density also varies across populations in the valley oak range [79] and showed strong associations with variation in SNPs and all three methylation sequence contexts. The associations with sequence contexts were substantially reduced after controlling for genetic variation, suggesting that variation in SNPs is likely to be the underlying driver of variation in trichome density. However, further studies, potentially using demethylating chemical agents to experimentally manipulate methylation levels [81], are needed to conclusively determine causal pathways and disentangle the influence of variation in DNA methylation levels from potential confounding effects of genomic background.

Given the complexities in the way that DNA methylation is associated with phenotypic variation depending on many factors, such as context of the methylation [7], TE insertions impacts (e.g., [45]), and variation in DNA methylation homeostasis (e.g., [16]), we acknowledge the limitations in this study. Once candidate phenotypes are identified, future experimental work would be necessary to identify the mechanisms underlying the patterns reported here.

In general, studies associating DNA methylation with phenotypic traits in non-model plant systems are rare, and the sample size of 160 individuals achieved in this study are high for current standards of ecological epigenomic studies. However, such a sample size is relatively low for human studies and studies in model plant systems [82]. As a consequence, our study was limited by low statistical power and high uncertainty in estimates of percent variance explained in phenotypic traits. Because of this limitation, we advise caution in interpreting the values of percent variance explained as exact estimates and recognize that the lack of a significant association between SNPs, DNA methylation, and a phenotypic trait could arise from low statistical power rather than a lack of a true association. In addition, detecting associations between polygenic traits and variation in SNPs or DNA methylation can be difficult, relying on adequate coverage of markers across the genome and the sequenced markers being in linkage disequilibrium with causal variants or potentially casual variants themselves [83]. From our simulation analysis (Figure 3, Figure S2), samples sizes on the order of 500–1000 would be needed to obtain ~75–100% statistical power depending on the strength of the association between DNA methylation and phenotypic trait. Achieving the sample sizes necessary to obtain sufficient statistical power and reduced uncertainty will become more feasible as epigenomic sequencing techniques continue to become more accessible in ecology and evolutionary biology research [4,70].

In summary, this study provides compelling evidence that DNA methylation can predict variation in ecologically relevant phenotypic traits, such as plant height, leaf lobedness, disease susceptibility, and trichome density in valley oak. Future studies that combine whole-genome bisulfite sequencing, gene expression analysis, TE insertion polymorphisms and phenotypic trait associations will provide a stronger and more detailed understanding of the potential mechanistic links between methylation at particular genomic regions and phenotypic traits.

Supplementary Materials: The following are available online at https://www.mdpi.com/article/10.3390/f12050569/s1, Figure S1: Results of PCA for SNP and DNA methylation data, Figure S2: Proportion of variance explained by simulated phenotypic and DNA methylation data.

Author Contributions: Conceptualization, L.B. and V.L.S.; methodology, L.B., B.M., S.F.-G., J.W.W., and V.L.S.; formal analysis, L.B.; resources, V.L.S. and J.W.W.; writing—original draft preparation, L.B.; writing—review and editing, L.B., B.M., S.F.-G., J.W.W., and V.L.S.; funding acquisition, V.L.S. and J.W.W. All authors have read and agreed to the published version of the manuscript.

Funding: This research was funded in part by a research award to VLS and colleagues from National Science Foundation Plant Genome Research Program (NSF IOS-#1444661). Post-doctoral funding for LB was provided by the La Kretz Center for California Conservation Science at UCLA and the NSF award.

Institutional Review Board Statement: Not applicable.

Informed Consent Statement: Not applicable.

Data Availability Statement: The data and code used in this study are available on Figshare: https://doi.org/10.6084/m9.figshare.14241884, accessed on 29 April 2021.

Acknowledgments: We acknowledge the native peoples of California as the traditional caretakers of the oak ecosystems sampled for this project. We thank Z. Celikkol, J. Basaran, C. Henriquez, D. Burge, C. Reimche, S. A. Swanson, M. Muia, A. Yoon, M. Hopkins, A. Copeland, K. Chang, A. Tiwari, C. Horn, V. Michaelis, X. Zhu, and K. Beckley for help with data collection. We are grateful for A. Delfino

Mix, A. Lentz, A. Albarran-Laras, P. Gugger, L. Crane, and many others for their help in establishing and maintaining the valley oak provenance trial. We thank E. Eskin, B. Pasaniuc, S. Sankararaman, and members of the Sork lab for their helpful comments and discussion. Genomic studies were supported by NSF PGRP grant (IOS-1444661) to V.L.S. Provenance trials were supported by the US Department of Agriculture Forest Service Pacific Southwest Research Station. Any use of product names is for informational purposes only and does not imply endorsement by the US Government.

Conflicts of Interest: The authors declare no conflict of interest.

References

1. Sow, M.D.; Allona, I.; Ambroise, C.; Conde, D.; Fichot, R.; Gribkova, S.; Jorge, V.; Le-Provost, G.; Pâques, L.; Plomion, C.; et al. Epigenetics in forest trees: State of the art and potential implications for breeding and management in a context of climate change. In *Advances in Botanical Research*; Mirouze, M., Bucher, E., Gallusci, P., Eds.; Academic Press: Cambridge, MA, USA, 2018; Volume 88, pp. 387–453.
2. Carbó, M.; Iturra, C.; Correia, B.; Colina, F.J.; Meijón, M.; Álvarez, J.M.; Cañal, M.J.; Hasbún, R.; Pinto, G.; Valledor, L. Epigenetics in forest trees: Keep calm and carry on. In *Epigenetics in Plants of Agronomic Importance: Fundamentals and Applications: Transcriptional Regulation and Chromatin Remodelling in Plants*; Alvarez-Venegas, R., De-la-Peña, C., Casas-Mollano, J.A., Eds.; Springer International Publishing: Cham, Switzerland, 2019; pp. 381–403. ISBN 9783030147600.
3. Balao, F.; Paun, O.; Alonso, C. Uncovering the contribution of epigenetics to plant phenotypic variation in Mediterranean ecosystems. *Plant Biol.* **2017**, 1–12. [CrossRef] [PubMed]
4. Richards, C.L.; Alonso, C.; Becker, C.; Bossdorf, O.; Bucher, E.; Colomé-Tatché, M.; Durka, W.; Engelhardt, J.; Gaspar, B.; Gogol-Döring, A.; et al. Ecological plant epigenetics: Evidence from model and non-model species, and the way forward. *Ecol. Lett.* **2017**, *20*, 1576–1590. [CrossRef]
5. Verhoeven, K.J.F.; VonHoldt, B.M.; Sork, V.L. Epigenetics in ecology and evolution: What we know and what we need to know. *Mol. Ecol.* **2016**, *25*, 1631–1638. [CrossRef] [PubMed]
6. Finnegan, E.J.; Genger, R.K.; Peacock, W.J.; Dennis, E.S. DNA methylation in plants. *Annu. Rev. Plant Physiol. Plant Mol. Biol.* **1998**, *49*, 223–247. [CrossRef] [PubMed]
7. Law, J.A.; Jacobsen, S.E. Establising, maintaining and modifying DNA methylation patterns in plants and animals. *Nat. Rev. Genet.* **2011**, *11*, 204 220. [CrossRef]
8. Bossdorf, O.; Richards, C.L.; Pigliucci, M. Epigenetics for ecologists. *Ecol. Lett.* **2008**, *11*, 106–115. [CrossRef] [PubMed]
9. Verhoeven, K.J.F.; Jansen, J.J.; van Dijk, P.J.; Biere, A. Stress-induced DNA methylation changes and their heritability in asexual dandelions. *New Phytol.* **2010**, *185*, 1108–1118. [CrossRef]
10. Zhang, Y.Y.; Fischer, M.; Colot, V.; Bossdorf, O. Epigenetic variation creates potential for evolution of plant phenotypic plasticity. *New Phytol.* **2013**, *197*, 314–322. [CrossRef]
11. Angers, B.; Castonguay, E.; Massicotte, R. Environmentally induced phenotypes and DNA methylation: How to deal with unpredictable conditions until the next generation and after. *Mol. Ecol.* **2010**, *19*, 1283–1295. [CrossRef]
12. Herrera, C.M.; Alonso, C.; Medrano, M.; Pérez, R.; Bazaga, P. Transgenerational epigenetics: Inheritance of global cytosine methylation and methylation-related epigenetic markers in the shrub *Lavandula latifolia*. *Am. J. Bot.* **2018**, *105*, 741–748. [CrossRef]
13. Herman, J.J.; Sultan, S.E. DNA methylation mediates genetic variation for adaptive transgenerational plasticity. *Proc. Royal Soc. B* **2016**, *283*, 20160988. [CrossRef] [PubMed]
14. Johannes, F.; Porcher, E.; Teixeira, F.K.; Saliba-colombani, V.; Albuisson, J.; Heredia, F.; Colot, V. Assessing the impact of transgenerational epigenetic variation on complex traits. *PLoS Genet.* **2009**, *5*, e1000530. [CrossRef] [PubMed]
15. Whipple, A.V.; Holeski, L.M. Epigenetic inheritance across the landscape. *Front. Genet.* **2016**, *7*, 189. [CrossRef]
16. Zhang, Y.; Wendte, J.M.; Ji, L.; Schmitz, R.J. Natural variation in DNA methylation homeostasis and the emergence of epialleles. *Proc. Natl. Acad. Sci. USA* **2020**, *117*, 4874–4884. [CrossRef]
17. Fieldes, M.A.; Amyot, L.M. Epigenetic control of early flowering in flax lines induced by 5- azacytidine applied to germinating seed. *J. Hered.* **1999**, *90*, 199–206. [CrossRef]
18. Dubin, M.J.; Zhang, P.; Meng, D.; Remigereau, M.S.; Osborne, E.J.; Casale, F.P.; Drewe, P.; Kahles, A.; Jean, G.; Vilhjálmsson, B.; et al. DNA methylation in Arabidopsis has a genetic basis and shows evidence of local adaptation. *Elife* **2015**, *4*, 1–23. [CrossRef] [PubMed]
19. Finnegan, E.J.; Peacock, W.J.; Dennis, E.S. Reduced DNA methylation in *Arabidopsis thaliana* results in abnormal plant development. *Proc. Natl. Acad. Sci. USA* **1996**, *93*, 8449–8454. [CrossRef]
20. Tatra, G.S.; Miranda, J.; Chinnappa, C.C.; Reid, D.M. Effect of light quality and 5-azacytidine on genomic methylation and stem elongation in two ecotypes of *Stellaria longipes*. *Physiol. Plant.* **2000**, *109*, 313–321. [CrossRef]
21. Bräutigam, K.; Vining, K.J.; Lafon-Placette, C.; Fossdal, C.G.; Mirouze, M.; Marcos, J.G.; Fluch, S.; Fraga, M.F.; Guevara, M.Á.; Abarca, D.; et al. Epigenetic regulation of adaptive responses of forest tree species to the environment. *Ecol. Evol.* **2013**, *3*, 399–415. [CrossRef]
22. Brockerhoff, E.G.; Barbaro, L.; Castagneyrol, B.; Forrester, D.I.; Gardiner, B.; González-Olabarria, J.R.; Lyver, P.O.; Meurisse, N.; Oxbrough, A.; Taki, H.; et al. Forest biodiversity, ecosystem functioning and the provision of ecosystem services. *Biodivers. Conserv.* **2017**, *26*, 3005–3035. [CrossRef]

23. Bonan, G.B. Forests and climate change: Forcings, feedbacks, and the climate benefits of forests. *Science* **2008**, *320*, 1444–1449. [CrossRef] [PubMed]

24. Beer, C.; Reichstein, M.; Tomelleri, E.; Ciais, P.; Jung, M.; Carvalhais, N.; Rödenbeck, C.; Arain, M.A.; Baldocchi, D.; Bonan, G.B.; et al. Terrestrial gross carbon dioxide uptake: Global distribution and covariation with climate. *Science* **2010**, *329*, 834–838. [CrossRef] [PubMed]

25. Cavender-Bares, J. Diversity, distribution and ecosystem services of the North American oaks. *Int. Oaks* **2016**, *27*, 37–48.

26. Platt, A.; Gugger, P.F.; Pellegrini, M.; Sork, V.L. Genome-wide signature of local adaptation linked to variable CpG methylation in oak populations. *Mol. Ecol.* **2015**, *24*, 3823–3830. [CrossRef]

27. Gugger, P.F.; Fitz-Gibbon, S.; Pellegrini, M.; Sork, V.L. Species-wide patterns of DNA methylation variation in *Quercus lobata* and their association with climate gradients. *Mol. Ecol.* **2016**, *25*, 1665–1680. [CrossRef] [PubMed]

28. Rico, L.; Ogaya, R.; Barbeta, A.; Peñuelas, J. Changes in DNA methylation fingerprint of *Quercus ilex* trees in response to experimental field drought simulating projected climate change. *Plant Biol.* **2014**, *16*, 419–427. [CrossRef] [PubMed]

29. Correia, B.; Valledor, L.; Meijón, M.; Rodriguez, J.L.; Dias, M.C.; Santos, C.; Cañal, M.J.; Rodriguez, R.; Pinto, G. Is the interplay between epigenetic markers related to the acclimation of cork oak plants to high temperatures? *PLoS ONE* **2013**, *8*, e53543. [CrossRef]

30. Ramos, M.; Rocheta, M.; Carvalho, L.; Inácio, V.; Graça, J.; Morais-Cecilio, L. Expression of DNA methyltransferases is involved in *Quercus suber* cork quality. *Tree Genet. Genomes* **2013**, *9*, 1481–1492. [CrossRef]

31. Browne, L.; Mead, A.; Horn, C.; Chang, K.; Celikkol, Z.A.; Henriquez, C.L.; Ma, F.; Beraut, E.; Meyer, R.S.; Sork, V.L. Experimental DNA demethylation associates with changes in growth and gene expression of oak tree seedlings. *G3* **2020**, *10*, 1019–1028. [CrossRef]

32. Yakovlev, I.A.; Asante, D.K.A.; Fossdal, C.G.; Junttila, O.; Johnsen, Ø. Differential gene expression related to an epigenetic memory affecting climatic adaptation in Norway spruce. *Plant Sci.* **2011**, *180*, 132–139. [CrossRef]

33. Liang, D.; Zhang, Z.; Wu, H.; Huang, C.; Shuai, P.; Ye, C.-Y.; Tang, S.; Wang, Y.; Yang, L.; Wang, J.; et al. Single-base-resolution methylomes of *Populus trichocarpa* reveal the association between DNA methylation and drought stress. *BMC Genet.* **2014**, *15*, S9. [CrossRef]

34. Jacinto Pereira, W.; de Castro Rodrigues Pappas, M.; Camargo Campoe, O.; Stape, J.L.; Grattapaglia, D.; Joannis Pappas, G., Jr. Patterns of DNA methylation changes in elite Eucalyptus clones across contrasting environments. *For. Ecol. Manag.* **2020**, *474*, 118319. [CrossRef]

35. Albaladejo, R.G.; Parejo-Farnés, C.; Rubio-Pérez, E.; Nora, S.; Aparicio, A. Linking DNA methylation with performance in a woody plant species. *Tree Genet. Genomes* **2019**, *15*, 15. [CrossRef]

36. Herrera, C.M.; Bazaga, P. Epigenetic correlates of plant phenotypic plasticity: DNA methylation differs between prickly and nonprickly leaves in heterophyllous *Ilex aquifolium* (Aquifoliaceae) trees. *Bot. J. Linn. Soc.* **2013**, *171*, 441–452. [CrossRef]

37. Kalisz, S.; Purugganan, M.D. Epialleles via DNA methylation: Consequences for plant evolution. *Trends Ecol. Evol.* **2004**, *19*, 309–314. [CrossRef]

38. Mirouze, M.; Paszkowski, J. Epigenetic contribution to stress adaptation in plants. *Curr. Opin. Plant Biol.* **2011**, *14*, 267–274. [CrossRef]

39. Dowen, R.H.; Pelizzola, M.; Schmitz, R.J.; Lister, R.; Dowen, J.M.; Nery, J.R.; Dixon, J.E.; Ecker, J.R. Widespread dynamic DNA methylation in response to biotic stress. *Proc. Natl. Acad. Sci. USA* **2012**, *109*, E2183–E2191. [CrossRef]

40. Niederhuth, C.E.; Schmitz, R.J. Putting DNA methylation in context: From genomes to gene expression in plants. *Biochim. Biophys. Acta* **2017**, *1860*, 149–156. [CrossRef]

41. Bewick, A.J.; Zhang, Y.; Wendte, J.M.; Zhang, X.; Schmitz, R.J. Evolutionary and experimental loss of gene body methylation and its consequence to gene expression. *G3* **2019**, *9*, 2441–2445. [CrossRef]

42. Kundariya, H.; Yang, X.; Morton, K.; Sanchez, R.; Axtell, M.J.; Hutton, S.F.; Fromm, M.; Mackenzie, S.A. MSH1-induced heritable enhanced growth vigor through grafting is associated with the RdDM pathway in plants. *Nat. Commun.* **2020**, *11*, 5343. [CrossRef]

43. Gienapp, P.; Fior, S.; Guillaume, F.; Lasky, J.R.; Sork, V.L.; Csilléry, K. Genomic quantitative genetics to study evolution in the wild. *Trends Ecol. Evol.* **2017**, *32*, 897–908. [CrossRef] [PubMed]

44. Yang, J.; Benyamin, B.; McEvoy, B.P.; Gordon, S.; Henders, A.K.; Nyholt, D.R.; Madden, P.A.; Heath, A.C.; Martin, N.G.; Montgomery, G.W.; et al. Common SNPs explain a large proportion of the heritability for human height. *Nat. Genet.* **2010**, *42*, 565–569. [CrossRef] [PubMed]

45. Domínguez, M.; Dugas, E.; Benchouaia, M.; Leduque, B.; Jiménez-Gómez, J.M.; Colot, V.; Quadrana, L. The impact of transposable elements on tomato diversity. *Nat. Commun.* **2020**, *11*, 4058. [CrossRef] [PubMed]

46. Thomson, C.E.; Winney, I.S.; Salles, O.C.; Pujol, B. A guide to using a multiple-matrix animal model to disentangle genetic and nongenetic causes of phenotypic variance. *PLoS ONE* **2018**, *13*, e0197720. [CrossRef] [PubMed]

47. Yang, J.; Lee, S.H.; Goddard, M.E.; Visscher, P.M. GCTA: A tool for genome-wide complex trait analysis. *Am. J. Hum. Genet.* **2011**, *88*, 76–82. [CrossRef]

48. Sork, V.L.; Cokus, S.J.; Fitz-Gibbon, S.T.; Zimin, A.V.; Puiu, D.; Garcia, J.A.; Gugger, P.F.; Henriquez, C.L.; Zhen, Y.; Lohmueller, K.E.; et al. High-quality genome and methylomes illustrate features underlying evolutionary success of oaks. *bioRxiv* **2021**. [CrossRef]

49. Niederhuth, C.E.; Bewick, A.J.; Ji, L.; Alabady, M.S.; Kim, K.D.; Li, Q.; Rohr, N.A.; Rambani, A.; Burke, J.M.; Udall, J.A.; et al. Widespread natural variation of DNA methylation within angiosperms. *Genome Biol.* **2016**, *17*, 194. [CrossRef]

50. Pavlik, B.M.; Muick, P.C.; Johnson, S.G.; Popper, M. *Oaks of California*; Cachuma Press: Los Olivos, CA, USA, 1991.

51. Anderson, M.K. *Indigenous Uses, Management, and Restoration of Oaks of the Far Western United States*; Natural Resources Conservation Service, National Plant Center, United States Department of Agriculture: Washington, DC, USA, 2007.

52. Gugger, P.F.; Ikegami, M.; Sork, V.L. Influence of late Quaternary climate change on present patterns of genetic variation in valley oak, *Quercus lobata* Nee. *Mol. Ecol.* **2013**, *22*, 3598–3612. [CrossRef]

53. Sork, V.L.; Davis, F.W.; Westfall, R.; Flint, A.; Ikegami, M.; Wang, H.; Grivet, D. Gene movement and genetic association with regional climate gradients in California valley oak (*Quercus lobata* Nee) in the face of climate change. *Mol. Ecol.* **2010**, *19*, 3806–3823. [CrossRef]

54. Delfino Mix, A.; Wright, J.W.; Gugger, P.F.; Liang, C.; Sork, V.L. Establishing a range-wide provenance test in valley oak (Quercus lobata Née) at two California sites. In *Proceedings of the Seventh Oak Symposium: Managing Oak Woodlands in a Dynamic World*; General Technical Report PSW-GTR-251; Standiford, R.B., Purcelll, K.L., Eds.; U.S. Department of Agriculture, Forest Service, Pacific Southwest Research Station: Berkeley, CA, USA, 2015; pp. 413–424.

55. MacDonald, B.W. Local Climatic Heterogeneity Predicts Differences in Phenotypic Plasticity across Populations of a Widely-Distributed California Oak Species. Master's Thesis, University of California, Los Angeles, CA, USA, 2017.

56. Sage, R.D.; Koenig, W.D.; McLaughlin, B.C. Fitness consequences of seed size in the valley oak *Quercus lobata* Née (Fagaceae). *Ann. For. Sci.* **2011**, *68*, 477. [CrossRef]

57. Rueden, C.T.; Schindelin, J.; Hiner, M.C.; DeZonia, B.E.; Walter, A.E.; Arena, E.T.; Eliceiri, K.W. ImageJ2: ImageJ for the next generation of scientific image data. *BMC Bioinform.* **2017**, *18*, 529. [CrossRef] [PubMed]

58. Sisó, S.; Camarero, J.; Gil-Pelegrín, E. Relationship between hydraulic resistance and leaf morphology in broadleaf Quercus species: A new interpretation of leaf lobation. *Trees* **2001**, *15*, 341–345. [CrossRef]

59. Pérez-Harguindeguy, N.; Díaz, S.; Lavorel, S.; Poorter, H.; Jaureguiberry, P.; Bret-Harte, M.S.; Cornwell, W.K.; Craine, J.M.; Gurvich, D.E.; Urcelay, C.; et al. New handbook for standardized measurment of plant functional traits worldwide. *Aust. J. Bot.* **2013**, *23*, 167–234. [CrossRef]

60. Niinemets, Ü. Global-scale climatic controls of leaf dry mass per area, density, and thickness in trees and shrubs. *Ecology* **2001**, *82*, 453–469. [CrossRef]

61. Gianfagna, T.J.; Carter, C.D.; Sacalis, J.N. Temperature and photoperiod influence trichome density and sesquiterpene content of *Lycopersicon hirsutum* f. Hirsutum. *Plant Physiol.* **1992**, *100*, 1403–1405. [CrossRef]

62. Picotte, J.J.; Rosenthal, D.M.; Rhode, J.M.; Cruzan, M.B. Plastic responses to temporal variation in moisture availability: Consequences for water use efficiency and plant performance. *Oecologia* **2007**, *153*, 821–832. [CrossRef]

63. Ning, P.; Wang, J.; Zhou, Y.; Gao, L.; Wang, J.; Gong, C. Adaptional evolution of trichome in *Caragana korshinskii* to natural drought stress on the Loess Plateau, China. *Ecol. Evol.* **2016**, *6*, 3786–3795. [CrossRef]

64. Bates, D.; Mächler, M.; Bolker, B.M.; Walker, S.C. Fitting linear mixed-effects models using lme4. *J. Stat. Softw.* **2015**, *67*, 1–48. [CrossRef]

65. Li, J.T.; Yang, J.; Chen, D.C.; Zhang, X.L.; Tang, Z.S. An optimized mini-preparation method to obtain high-quality genomic DNA from mature leaves of sunflower. *Genet. Mol. Res.* **2007**, *6*, 1064–1071.

66. Feng, S.; Rubbi, L.; Jacobsen, S.E.; Pellegrini, M. Determining DNA methylation profiles using sequencing. *Methods Mol. Biol.* **2011**, *733*, 223–238. [CrossRef]

67. Pedersen, B.S.; Eyring, K.; De, S.; Yang, I.V.; Schwartz, D.A. Fast and accurate alignment of long bisulfite-seq reads. *arXiv* **2014**, arXiv:1401.1129.

68. Sork, V.L.; Cokus, S.J.; Fitz-Gibbon, S.T.; Zimin, A.; Puiu, D.; Pellegrini, M.; Salzberg, S.L. Valley Oak Genomic Resources. Available online: https://valleyoak.ucla.edu/genomic-resources/ (accessed on 31 October 2019).

69. Andri, S. DescTools: Tools for Descriptive Statistics. Available online: https://cran.r-project.org/web/packages/DescTools/index.html (accessed on 31 October 2019).

70. Lea, A.J.; Vilgalys, T.P.; Durst, P.A.P.; Tung, J. Maximizing ecological and evolutionary insight in bisulfite sequencing data sets. *Nat. Ecol. Evol.* **2017**, *1*, 1074–1083. [CrossRef] [PubMed]

71. McKenna, A.; Hanna, M.; Banks, E.; Sivachenko, A.; Cibulskis, K.; Kernytsky, A.; Garimella, K.; Altshuler, D.; Gabriel, S.; Daly, M.; et al. The Genome Analysis Toolkit: A Map Reduce framework for analyzing next-generation DNA sequencing data. *Genome Res.* **2010**, *20*, 1297–1303. [CrossRef] [PubMed]

72. Danecek, P.; Auton, A.; Abecasis, G.; Albers, C.A.; Banks, E.; DePristo, M.A.; Handsaker, R.E.; Lunter, G.; Marth, G.T.; Sherry, S.T.; et al. The variant call format and VCFtools. *Bioinformatics* **2011**, *27*, 2156–2158. [CrossRef] [PubMed]

73. Purcell, S.; Neale, B.; Todd-Brown, K.; Thomas, L.; Ferreira, M.A.R.; Bender, D.; Maller, J.; Sklar, P.; de Bakker, P.I.W.; Daly, M.J.; et al. PLINK: A tool set for whole-genome association and population-based linkage analyses. *Am. J. Hum. Genet.* **2007**, *81*, 559–575. [CrossRef] [PubMed]

74. Zhang, F.; Chen, W.; Zhu, Z.; Zhang, Q.; Nabais, M.F.; Qi, T.; Deary, I.J.; Wray, N.R.; Visscher, P.M.; McRae, A.F.; et al. OSCA: A tool for omic-data-based complex trait analysis. *Genome Biol.* **2019**, *20*, 107. [CrossRef]

75. Main, B.J.; Lee, Y.; Ferguson, H.M.; Kreppel, K.S.; Kihonda, A.; Govella, N.J.; Collier, T.C.; Cornel, A.J.; Eskin, E.; Kang, E.Y.; et al. The genetic basis of host preference and resting behavior in the major African malaria vector, *Anopheles arabiensis*. *PLoS Genet.* **2016**, *12*, e1006303. [CrossRef]

76. Browne, L.; Wright, J.W.; Fitz-Gibbon, S.; Gugger, P.F.; Sork, V.L. Adaptational lag to temperature in valley oak (*Quercus lobata*) can be mitigated by genome-informed assisted gene flow. *Proc. Natl. Acad. Sci. USA* **2019**, *116*, 25179–25185. [CrossRef]

77. Bossdorf, O.; Arcuri, D.; Richards, C.L.; Pigliucci, M. Experimental alteration of DNA methylation affects the phenotypic plasticity of ecologically relevant traits in *Arabidopsis thaliana*. *Evol. Ecol.* **2010**, *24*, 541–553. [CrossRef]

78. Kanchanaketu, T.; Hongtrakul, V. Treatment of 5-azacytidine as DNA demethylating agent in *Jatropha curcas* L. *Kasetsart J-Nat. Sci.* **2015**, *49*, 524–535.

79. MacDonald, B.W.S.; Wright, J.W.; Sork, V.L. *Local climatic Heterogeneity Predicts Differences in Phenotypic Plasticity across Populations of a Widely-Distributed California Oak, Quercus lobata Nee*; In preparation.

80. Ramirez, H.-L.; Ivey, C.T.; Wright, J.W.; MacDonald, B.W.S.; Sork, V.L. Variation in leaf shape in a *Quercus lobata* common garden: Tests for adaptation to climate and physiological consequences. *Madrono* **2020**, *67*, 77–84. [CrossRef]

81. Griffin, P.T.; Niederhuth, C.E.; Schmitz, R.J. A comparative analysis of 5-Azacytidine- and zebularine-induced DNA demethylation. *G3* **2016**, *6*, 2773–2780. [CrossRef] [PubMed]

82. Visscher, P.M.; Hemani, G.; Vinkhuyzen, A.A.E.; Chen, G.-B.; Lee, S.H.; Wray, N.R.; Goddard, M.E.; Yang, J. Statistical power to detect genetic (co)variance of complex traits using SNP data in unrelated samples. *PLoS Genet.* **2014**, *10*, e1004269. [CrossRef] [PubMed]

83. Wray, N.R.; Yang, J.; Hayes, B.J.; Price, A.L.; Goddard, M.E.; Visscher, P.M. Pitfalls of predicting complex traits from SNPs. *Nat. Rev. Genet.* **2013**, *14*, 507–515. [CrossRef] [PubMed]

*forests*

Article

Genetic, Morphological, and Environmental Differentiation of an Arid-Adapted Oak with a Disjunct Distribution

Bethany A. Zumwalde [1,2,*], Ross A. McCauley [3], Ian J. Fullinwider [3], Drew Duckett [1,4], Emma Spence [1] and Sean Hoban [1,*]

[1] The Morton Arboretum, Center for Tree Science, 4100 Illinois 53, Lisle, IL 60532, USA; duckett.17@osu.edu (D.D.); emma@largelandscapes.org (E.S.)
[2] Department of Biology, University of Florida, Gainesville, FL 32611, USA
[3] Department of Biology, Fort Lewis College, 1000 Rim Drive, Durango, CO 81301, USA; McCauley_R@fortlewis.edu (R.A.M.); ijfullinwider@fortlewis.edu (I.J.F.)
[4] Department of Evolution, Ecology and Organismal Biology, The Ohio State University, 1315 Kinnear Rd., Columbus, OH 43212, USA
* Correspondence: bzumwalde@ufl.edu (B.A.Z.); shoban@mortonarb.org (S.H.)

Abstract: The patterns of genetic and morphological diversity of a widespread species can be influenced by environmental heterogeneity and the degree of connectivity across its geographic distribution. Here, we studied *Quercus havardii* Rydb., a uniquely adapted desert oak endemic to the Southwest region of the United States, using genetic, morphometric, and environmental datasets over various geographic scales to quantify differentiation and understand forces influencing population divergence. First, we quantified variation by analyzing 10 eastern and 13 western populations from the disjunct distribution of *Q. havardii* using 11 microsatellite loci, 17 morphological variables, and 19 bioclimatic variables. We then used regressions to examine local and regional correlations of climate with genetic variation. We found strong genetic, morphological and environmental differences corresponding with the large-scale disjunction of populations. Additionally, western populations had higher genetic diversity and lower relatedness than eastern populations. Levels of genetic variation in the eastern populations were found to be primarily associated with precipitation seasonality, while levels of genetic variation in western populations were associated with lower daily temperature fluctuations and higher winter precipitation. Finally, we found little to no observed environmental niche overlap between regions. Our results suggest that eastern and western populations likely represent two distinct taxonomic entities, each associated with a unique set of climatic variables potentially influencing local patterns of diversity.

Keywords: *Quercus havardii*; Fagaceae; genetic differentiation; morphometrics; bioclimatic associations

Citation: Zumwalde, B.A.; McCauley, R.A.; Fullinwider, I.J.; Duckett, D.; Spence, E.; Hoban, S. Genetic, Morphological, and Environmental Differentiation of an Arid-Adapted Oak with a Disjunct Distribution. *Forests* **2021**, *12*, 465. https://doi.org/10.3390/f12040465

Academic Editors: Mary Ashley and Janet R. Backs

Received: 20 February 2021
Accepted: 9 April 2021
Published: 10 April 2021

1. Introduction

Genetic and phenotypic trait diversity of populations and species may be strongly influenced by environmental conditions at regional and local scales [1,2]. For example, favorable environmental conditions may promote higher genetic diversity by influencing the number of populations and migrants they produce, and thus increase their genetic connectivity [3]. Alternatively, unfavorable conditions could reduce genetic diversity and increase genetic drift by influencing population bottlenecks, inbreeding among close relatives, or local extinction–colonization dynamics [4]. Furthermore, environmental conditions can impose selective pressures and influence the rate at which populations diverge and speciation may occur [5–7].

An additional factor influencing genetic and trait divergence is large-scale spatial disjunctions, or discontinuities, in the range across which little or no migration occurs. Species with a once continuous range, but now consisting of disjunct populations, have long been of biogeographical interest [8–10], especially when those disjunctions have been caused by climatic changes from events such as Pleistocene glacial cycles [11,12]. Disjunct distributions resulting in spatial isolation can influence the divergence of populations or ultimately result in parapatric speciation. This isolation may enable genetic changes that determine distinctive morphological and physiological characters [13].

Studies in *Quercus* L. (Fagaceae), one of the most economically and ecologically important woody genera of northern temperate zones, have been particularly informative for examining environmental [14,15], genetic [16,17], and trait-based [18,19] relationships in regard to geographic and biogeographic history [20,21]. Species in this genus have adapted to a remarkable range of habitats [22,23]; however, relatively few oaks in the United States are adapted to the arid environments of the Southwest [24,25], and in particular, to habitats with deep, shifting sand. A key research goal in understanding adaptation in desert oaks is to identify how this extreme environment has influenced genetic and morphological diversity, to better understand the range of conditions that oaks can survive in, and to apply this information to conserving oaks in situ.

To investigate the degree to which the environment contributes to genetic and phenotypic diversity, we examine Havard's oak, *Quercus havardii* Rydb., a keystone species that defines the sand shinnery ecosystem [26,27]. Commonly known as shinnery oak, *Q. havardii* is often found in harsh, arid environments as thickets or large shrubs in deep sands, which are not typically shared by other species of oak. This environment is quite extreme, and this growth habit is unusual among oaks. It has been reported as being highly clonal and rarely producing sexually via acorns [26,28,29]. As part of a recent evaluation of US oaks following IUCN Red List criteria [30], *Q. havardii* was listed as "Endangered" due to the increasing loss of habitat from agricultural practice and oil drilling, decreasing population size projections, and the total number of mature individuals.

Additionally, *Q. havardii* has a disjunct range: an eastern region ranging from western Oklahoma through to the Llano Estacado of west Texas and eastern New Mexico, and a western region encompassing the Navajo Basin in northeastern Arizona and southeastern Utah [31]. Few populations occur in the broad disjunction in central New Mexico, although the presence of this species in this area has been theorized to represent a continuous distribution in the past [32]. There has been no work to date examining the morphological and genetic diversity and structure of *Q. havardii* across the entirety of its range, nor any work on the range-wide population structure of any of the southwestern United States oaks (e.g., *Q. gambelii* Nutt., *Q. turbinella* Greene, *Q. ajoensis* C.H. Muller). Thus, we lack general knowledge of how oak populations are influenced by the harsh environmental conditions of desert habitats.

In this study, we aim to quantify genetic, morphological and environmental differentiation in populations of *Q. havardii*. We then summarize differentiation at both local and regional scales, and assess the timing at which population divergence potentially occurred. Finally, we evaluate how key environmental factors (i.e., precipitation and temperature) possibly contribute to this genetic variation within and across populations.

2. Materials and Methods

2.1. Sample Collection

Leaf tissue was collected in the summer of 2016 from a total of 489 samples from 23 populations for both morphological and genetic studies (Figure 1). From each individual, two leaves were collected and pressed for morphometric analysis and two to three leaves were placed on ice for DNA extraction. Heteroblastic leaf shape has been noted to occur in Fagaceae [33,34], in which leaf morphology may differ from various shoot types. Using a collection method implemented in similar morphometric studies of *Quercus* [32,35,36], we collected fully expanded and undamaged leaves from sun-exposed terminal shoots. A total of 23 populations were included in genetic analysis, while only 21 populations were included in morphological analysis. The two populations that were not included in the morphological analysis (E13 and E14) were sampled by collaborators and only included enough tissue for use in genetic analysis. Voucher specimens were produced for each population and deposited at MOR and FLD with vouchers collected from the Navajo Nation additionally deposited in NAVA (herbarium acronyms follow Holmgren et al. [37]). Populations were chosen by considering timing and logistical issues of sampling while also attempting to represent the geographic range of the taxon. A list of possible sites was generated by contacting land managers of private and public land in the region, consulting GBIF (www.gbif.org; accessed 14 March 2015) and SEINet (https://swbiodiversity.org/seinet; accessed on 14 March 2015), and using suggestions from members of the International Oak Society. Additionally, populations were selected that were subject to little or no observed hybridization, although two sites located within the major disjunction (W11 and W12) documented occurrences of *Q. gambelii* and *Q. turbinella* in the vicinity.

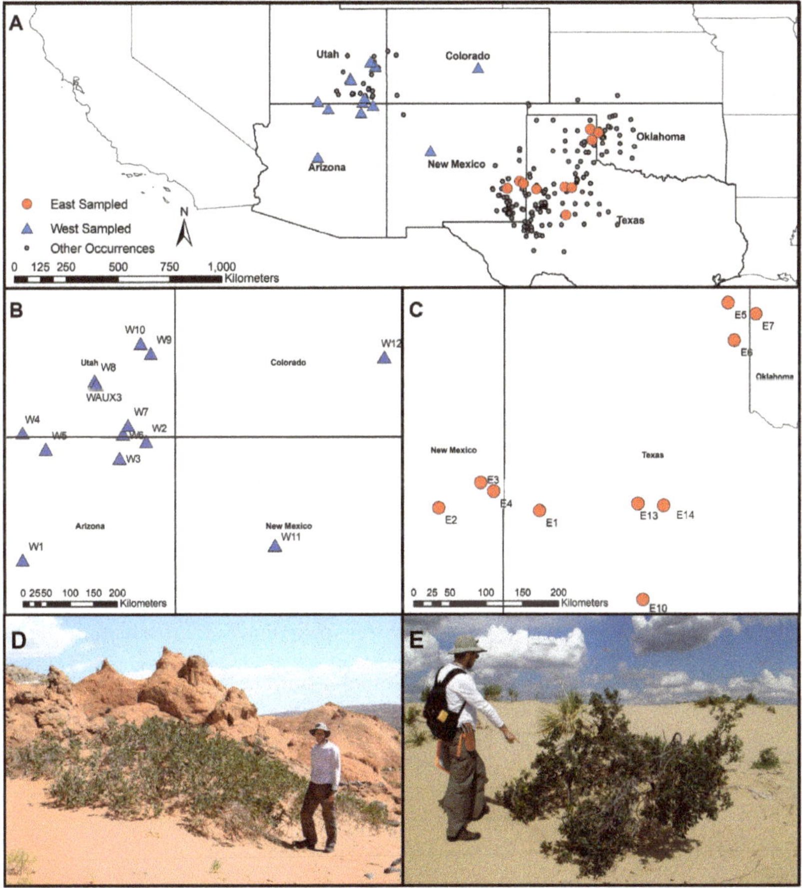

Figure 1. Sampling and distribution maps of *Quercus havardii*. (**A**) Known occurrence points for *Q. havardii* in grey circles and sampled populations used for genetic, morphological, and environmental analysis in blue triangles (western range) and red circles (eastern range); (**B**) sampling sites of 13 populations from the western range; (**C**) sampling sites of 10 populations from the eastern range; (**D**) example of *Q. havardii* habit from western region population W8; (**E**) example of *Q. havardii* in eastern region population E2.

2.2. DNA Extraction and Microsatellite Protocol

Genetic diversity was evaluated based on 11 nuclear microsatellite loci developed for other *Quercus* species (Appendix A Table A1). Genomic DNA was extracted from frozen leaf tissue using the E.Z.N.A. Plant DNA DS Kit (Omega Bio-tek, Norcross, GA, USA) and diluted to an approximate concentration of 5 ng/µL for PCR amplification. Microsatellite primers were organized using the software Multiplex Manager v.1.2 [38]. Multiplexed PCRs were carried out in a total volume of 10 µL, containing 2 µL of DNA template (5 ng/µL) and 8 µL of master mix consisting of 1× reaction buffer, 0.5 mM total dNTPs, 1.5 mM $MgCl_2$, fluorescently tagged primer concentrations ranging from 0.02 to 0.28 µM, 0.03 µM of each reverse primer, 0.5 µg/µL BSA, 0.025 U of GoTaq G2 DNA polymerase (Promega, Madison, WI, USA) and HPLC-grade water. Initial PCR conditions and reagents, including the use of BSA, were based upon similar microsatellite studies of congeneric species [39–42]. Forward primers were labeled with one of the 6-FAM, NED, VIC or PET fluorescent dyes (Applied Biosystems, Waltham, MA, USA). Cycling conditions were: an initial denaturation step at 94 °C for 5 min, then 30 cycles consisting of 30 s at 94 °C, 30 s at 50–56 °C (multiplex 1 at 50 °C, multiplex 2 at 54 °C, and multiplex 3 at 56 °C) and 30 s at 72 °C, followed by a final extension of 5 min at 72 °C. PCRs were ran on Eppendorf Mastercycler pro (Eppendorf, Hamburg, Germany) and C-1000 Touch (Bio-Rad, Hercules, CA, USA) machines. Amplification products were visualized using a 1.5% agarose, 1× TAE gel with ethidium bromide stain and sized with a 1 kb + ladder (Thermofisher Scientific, Waltham, MA, USA). PCR products were run on an ABI 3730 Genetic Capillary Electrophoresis Sequencer (Applied Biosystems) and fragment sizes of alleles were scored using the software Geneious v.10.2.3 (Biomatters, Auckland, New Zealand), using the microsatellite plugin v.1.4.4. Alleles called and binned by hand were compared with suggested bins from Geneious. Micro-Checker v.2.2.3 [43] and INESTv.2.2 [44] were both used to check for null alleles across loci.

2.3. Statistical Analysis of Genetic Data

Initial analyses were performed using a minimum of 10 individuals per population, which means that some initially sampled locations were not analyzed due to having too few individuals (E11, E13, and W2). We used R v.3.6.3 [45] for all R-based analyses. After the removal of clones from the dataset using the R package Poppr v.2.9.0 [46], we decided to decrease the minimum number of individuals per population to seven in order to maximize the number of populations included over the range of the taxon. The presence of null alleles was detected in several loci. Thus, analyses were performed both including and discarding these loci. The final dataset reported in this study used genetic data from eleven microsatellite loci for a total of 489 individuals from 23 populations, requiring a minimum of seven individuals per population (Table 1). Alternative analyses discarding loci with null alleles and adjusting minimum population requirements were performed and did not alter results; these can be run using the code on GitHub (doi:10.5281/zenodo.4453098). Genetic diversity statistics including allelic richness, heterozygosity, and F_{ST} were calculated using the R packages adegenet v.2.1.3 [47], hierfstat v.0.5-7 [48], and demerlate v.0.9-3 [49]. Tests for normality were performed using the Shapiro–Wilk test for normality and Q-Q plots, which showed no significant departures from normality for the three diversity and relatedness measures, but did show a significant non-normal distribution for F_{ST}. This is true for all subsets of the data tested. To compare populations from each region (East vs. West) for these basic summary statistics, we performed *t*-tests, as well as Wilcoxon tests, for these four statistics, for all subsets of the data. To determine patterns of isolation by distance and the distribution of genetic variation between and among populations, a Mantel Test and an Analysis of Molecular Variance (AMOVA; F_{ST}, 999 permutations) were performed using GenAlEX v.6.503 [50]. Genetic data were also visualized using principal coordinate analysis (PCoA) of pairwise distances between populations using GenAlEx. Genetic structure was visualized using STRUCTURE v.2.3.4 software [26,27]. STRUCTURE plots were examined for both range-wide and within regions [51]. K values were set from

1 to 40 with 20 iterations at 100,000 burn-ins and Markov Chain Monte Carlo (MCMC) repeats. Mean likelihood and ΔK values were obtained using Structure Harvester [52]. CLUMPP version 1.1.2 was used to resolve multimodality and DISTRUCT version 1.1 [53] was used to visualize results from STRUCTURE outputs. STRUCTURE clustering results were cross-checked with the model-free method discriminant analysis of principal components (DAPC) using the R package adegenet [47,54]. The number of K groups that best fit the data was chosen using the Evanno method [55].

Table 1. Population information and genetic diversity indices of the 23 populations of *Quercus havardii* using 11 microsatellite loci.

Population	Latitude (N)	Longitude (W)	*n*	*MLG*	*Clones*	*He*	*Ho*	*Na*	*Ar*	*F$_{ST}$*	*Rel*
E1	33.373	−102.617	20	18	2	0.554	0.524	58	3.61	0.088	0.215
E2	33.408	−103.867	25	22	3	0.562	0.496	56	3.42	0.090	0.250
E3	33.712	−103.344	27	20	7	0.520	0.418	57	3.44	0.094	0.239
E4	33.609	−103.183	19	13	6	0.542	0.433	49	3.58	0.082	0.149
E5	35.903	−100.276	32	17	15	0.543	0.516	60	3.55	0.095	0.265
E6	35.445	−100.199	14	11	3	0.654	0.570	60	4.26	0.078	0.113
E7	35.767	−99.926	56	40	16	0.615	0.554	91	4.16	0.072	0.156
E10	32.286	−101.332	9	9	0	0.649	0.581	55	4.38	0.066	0.054
E13	33.459	−101.391	8	8	0	0.634	0.648	53	4.29	0.073	0.158
E15	33.435	−101.074	24	21	3	0.704	0.591	84	4.82	0.061	0.009
W1	34.727	−111.985	39	25	14	0.553	0.478	58	3.37	0.129	0.275
W2	36.927	−109.618	11	7	4	0.681	0.494	53	4.57	0.077	−0.044
W3	36.616	−110.133	33	32	1	0.586	0.491	88	4.53	0.083	0.098
W4	37.096	−111.984	33	26	7	0.646	0.614	79	4.42	0.076	0.143
W5	36.786	−111.540	30	27	3	0.668	0.616	103	5.05	0.061	0.052
W6	37.064	−110.069	33	31	2	0.663	0.560	106	4.90	0.061	0.051
W7	37.234	−109.973	32	30	2	0.660	0.567	93	4.77	0.064	0.078
W8	38.065	−110.602	39	35	4	0.716	0.616	98	4.94	0.060	0.035
W9	38.573	−109.527	30	28	2	0.701	0.552	106	5.23	0.058	−0.001
W10	38.762	109.725	29	22	7	0.696	0.603	90	5.14	0.066	0.021
W11	34.998	−107.172	20	19	1	0.647	0.584	84	4.80	0.086	0.089
W12	38.502	−105.106	18	15	3	0.601	0.453	69	4.38	0.093	0.081
WAUX3	38.000	−110.569	16	14	2	0.694	0.641	73	4.86	0.070	0.051
East Mean			23.4	17.9	5.5	0.598	0.533	62.3	3.95	0.080	0.161
West Mean			27.9	23.9	4.0	0.655	0.559	84.6	4.69	0.076	0.072
Overall Mean			26.0	21.3	4.7	0.630	0.548	74.1	4.79	0.077	0.110

n: sample size; *MLG*: No. multilocus genotype; *Clones*: No. sampled clones in the population; *He*: expected heterozygosity; *Na*: No. alleles; *Ar*: allelic richness by locus; *F$_{ST}$*: pairwise genetic differentiation; *Rel*: relatedness within a population.

2.4. Morphometric Methods

We used morphometric analysis, utilizing the characters defined in McCauley et al. [32], to assess patterns of morphological differentiation at local and regional scales (Table 2). Twelve continuous characters based on defined landmarks from 2 leaves per individual were measured manually on a total of 928 leaves. These characters included the number of lobes (LOBES), total leaf length (LENGTH), basal lobe blade width (BLBW), middle lobe blade width (MLBW), apical lobe blade width (ALBW), lower vein length (LVL), upper vein length (UVL), upper middle lobe sinus width (UMLS), lower middle lobe sinus width (LMLS), angle of lower lobe (LLA), angle of upper lobe (ULA), and angle of middle lobe from apex (MLA). Additionally, five composite variables derived from the principal measurements based on ratios of leaf length to width (LWR) and lobe depth (LODR) were included: LWR_1 (BLBW/Length), LWR_2 (MLBW/Length), LWR_3 (ALBW/Length), LODR_1 (MLBW-UMLS/MLBW), LODR_2 (MLBW-LMLS/MLBW). While specifically adapted for southwestern oaks, these measurements represent typical landmarks used in morphometric analysis in *Quercus* [56,57]. Normality of data was tested both across and within populations for individual characters using a Shapiro–Wilk test. As the data for many characters were not normally distributed, we used non-parametric Spearman's correlation to identify potential correlation ($r > 0.70$) [58,59] among variables. As no correlation was observed, all variables were used in the subsequent analyses. Principal component analysis (PCA) was used to visualize differentiation of all individual samples and means from each population across the two disjunct regions with the first and second principal components plotted using the R package ggplot2 v.3.3.3 [60]. To explore the partitioning of morphological variance in each of the measured variables across multiple levels, we performed a linear mixed effects model using the R package nlme v.3.1-152 [61]. Region was considered a fixed effect in the model, while population within each region and trees within each population were considered random effects. Residual plots were used to check for normality. The proportion of the total variance contributed by the fixed and combined random effects was determined using the marginal and conditional R^2 procedure of Nakagawa and Schielzeth [62] and calculated using the R package MuMIn v. 1.43.4 [63].

Table 2. Proportion of morphological variance in 12 directly measured and five composite variables in *Q. havardii* partitioned by East and West (regional) and by combined population and individual (local).

Character	Abbreviation	Regional	Local
Lobe Number	LOBES	0.182	0.345
Leaf Length	LENGTH	0.309	0.332
Basal Lobe Blade Width	BLBW	0.064	0.449
Middle Lobe Blade Width	MLBW	0.010	0.491
Apical Lobe Blade Width	ALBW	0.023	0.025
Lower Vein Length	LVL	0.105	0.145
Upper Vein Length	UVL	0.002	0.342
Upper Middle Lobe Sinus Width	UMLS	0.001	0.508
Lower Middle Lobe Sinus Width	LMLS	0.027	0.470
Lower Lobe Angle	LLA	0.383	0.202
Upper Lobe Angle	ULA	0.038	0.161
Middle Lobe Angle	MLA	0.038	0.205
Length to Width Ratio 1 (BLBW/Length)	LWR_1	0.058	0.442
Length to Width Ratio 2 (MLBW/Length)	LWR_2	0.403	0.264
Length to Width Ratio 3 (ALBW/Length)	LWR_3	0.040	0.223
Lobe Depth Ratio 1	LODR_1	0.026	0.402
Lobe Depth Ratio 2	LODR_2	0.076	0.411

2.5. Environmental Differentiation and Influence on Genetic Variation

We used PCA and calculated means of bioclimatic variables to investigate and visualize environmental relationships among our sampled populations. We utilized the 19 bioclimatic variables from the WorldClim version 2.1 database [64] and variables were downloaded at raster grids of 2.5 arc minutes (ca. 5 km) to compensate for the total area of the largest populations sampled. The full names and abbreviations for bioclimatic variables used for each environmental analysis are listed in Table A6. To avoid multicollinearity, one of each pair of highly correlated variables ($r > 0.7$) was removed [58], leaving 6 bioclimatic variables for analysis. Of these remaining variables, 4 were associated with temperature and 2 with precipitation. PCA was used to visualize climatic similarities and dissimilarities of the 23 sampled populations using the R package ggplot2.

In an additional analysis examining environmental differences between the regions, we assessed environmental niche overlap. To do so, we used a dataset that consisted of 303 occurrence points (243 from the East and 60 from the West) after accounting for spatial autocorrelation. Occurrence points were provided by collaborators who produced the IUCN Red List assessment for oaks in the United States [30]. The 19 bioclimatic variables were downloaded for 2.5 arc minutes resolution. Highly correlated variables ($r > 0.7$) were removed, leaving 7 bioclimatic variables for analysis, of which 4 were associated with temperature and 2 were associated with precipitation (Table A6). Niche differentiation in environmental space and also niche equivalency, similarity, and overlap were calculated using the R package ecospat v3.2 [65]. Niche overlap was measured by Schoener's D index, which ranges from 0 (no overlap) to 1 (full overlap; [66]).

To visualize specific environmental and genetic (i.e., heterozygosity, allelic richness, and relatedness) relationships within and among regions, we built models by multiple linear regressions for 3 genetic statistics (heterozygosity, allelic richness, and relatedness) using uncorrelated environmental variables at several geographic scales. Given the disjunct and patchy distribution of this species, and confirmation that different variables were found to be correlated within each region, we used three different sets of uncorrelated bioclimatic variables corresponding to: (1) the range-wide distribution including both eastern and western sampled populations; (2) eastern populations only; (3) western populations only. Spearman correlation analyses were performed for every subgroup and correlated variables were removed ($r > 0.7$). A final list of uncorrelated variables used as inputs for each model can be found in Table A6. The variables selected across all subgroups were subjected to a final check for collinearity using variance inflation factor (VIF) ≤ 10 [58]. We compared regression models using the all-subsets variable reduction approach (R function regsubsets, package leaps v.3.1 [67]) and selected the "best supported" model using Bayes' BIC. Model performance was assessed using adjusted coefficients of determination (adjusted R^2) and the Benjimini and Hochberg [68] multiple comparisons correction for p-values.

2.6. Estimation of Divergence between East and West Regions Using ABC and Genetic Data

Modeling-based simulation analyses, such as approximate Bayesian computation (ABC), have been useful in elucidating the evolutionary histories of species [69–71]. We performed an ABC analysis to estimate the timing of divergence between the East and West regions. We compared two simple hypotheses: (1) splitting and isolation (the I model) or (2) splitting and continued migration (the IM model). We estimated the following parameters: splitting time (TS), mutation rate-scaled population sizes of each region, theta 1 and theta 2 (representing the East and West regions, respectively), and migration (M). We used the R code developed by Navascués [72], which runs simulations in the coalescent simulator ms [73] and uses random forest [74,75] for parameter estimation. This software simulates parameters in the form of theta, M and TS, which are scaled parameters rather than the parameters of biological interest here (i.e., population size and migration). An estimation of mutation rate is needed to transform these to population size, migration rate and splitting time on scales amendable to biological interpretation. We searched the literature for plant microsatellite mutation rates per site per generation and found four: wheat (2.4×10^{-4}), maize (7.7×10^{-4}), *Arabidopsis* (8.87×10^{-4}), and red cedar (6.3×10^{-4}) [76–79]. We then used the mean of the four values, 6.3×10^{-4}.

3. Results

3.1. Genetic Variation and Differentiation

Genetic analyses showed an overall pattern of differentiation between the East and West regions of *Q. havardii*. Summary statistics for all populations as well as means and standard deviation across populations are shown in Table 1. Genetic diversity was higher in western populations ($He = 0.655 \pm 0.04$) than eastern populations ($He = 0.598 \pm 0.061$). Using a Mantel Test (Figure A1), isolation by distance was significant when comparing among regions, as well as in the West ($R^2 = 0.256$, $p < 0.001$), but not in the East. The results of the AMOVA show that most of the molecular variance (81%) was found within and among individuals; however, more genetic variance occurred among regions (15%) than among populations (4%). When comparing between regions (Table A2), there is significantly higher allelic richness ($p = 0.0003$), higher heterozygosity ($p = 0.015$), and lower relatedness in West populations ($p = 0.009$; Figure 2) than in East populations. With the exception of population W1, which had significantly higher clones than other western populations, we detected a higher frequency of clones within eastern populations. Patterns from the first three axes of separate PCoA analysis among individuals (Figure A2) and populations (Figure 3A) revealed genetic distinction between the two regions and subdivisions of populations within the West, with W1 being the most separated from other populations. Bayesian clustering analyses from STRUCTURE (Figure 3D–F) and a discriminant analysis of principal components using the R package adegenet showed highly similar relationships among populations. For all three analyses (i.e., range-wide, eastern region, and western region), K = 2 was the best-supported assumption for each dataset. The next best supported number of clusters (K = 3) for the range-wide dataset is also shown (Figure 3D).

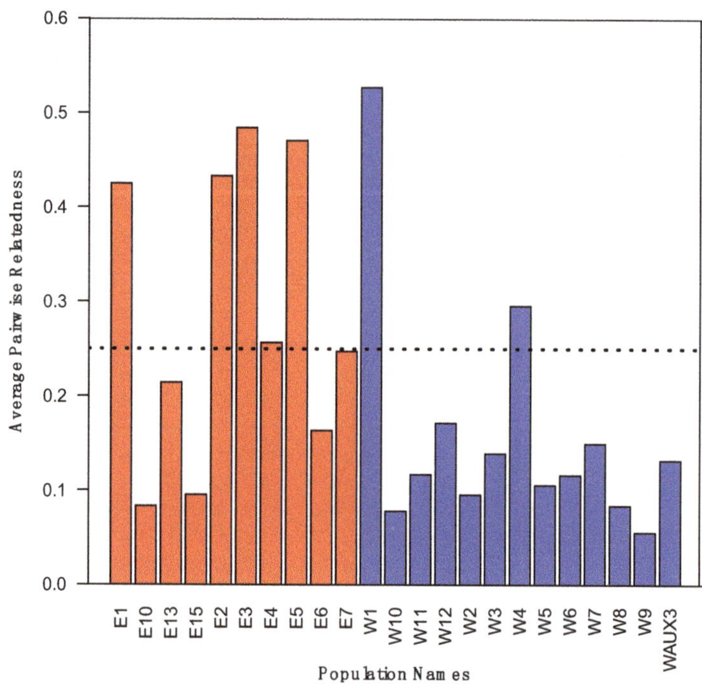

Figure 2. Mean pairwise relatedness within each *Quercus havardii* population, with blue and red indicating western and eastern populations, respectively. The expected value of relatedness for full siblings is shown by the horizontal dotted red line.

3.2. Morphological Variation

Similarly, the morphological analysis showed the presence of two groupings corresponding to the two regions. This was most pronounced when using population means, which indicated a level of differentiation similar to that seen in the genetic analysis (Figure 3B). The first two principal components represented 62.5% of the variation (Table A3; PCA plot loading scores are given in Table A4). Examination of population means showed most individual populations to cluster relatively close together in the West but with more dispersion of populations in PCA space than in the East (Figures A3 and A4). One western population, W4, was found to be highly divergent from the remaining populations. However, analysis at the individual-level (Figures A5 and A6) showed a high degree of morphological variation occurring across both regions and interdigitation between the East and West regions with a high degree of variation seen within most populations. The partitioning of morphological variance indicated that factors associated with leaf length and lower lobe angle (Table A5) contributed the most to separation between the two regions while high levels of variation at population- and tree-levels, strongest for factors associated with leaf width, indicated high levels of within-population variation (Table 2). Generally, leaves in the East were characterized by having longer leaves with slightly fewer lobes and more acute lobes, and leaves in the West were characterized by having shorter leaves that are widest at the middle lobe of the leaf blade.

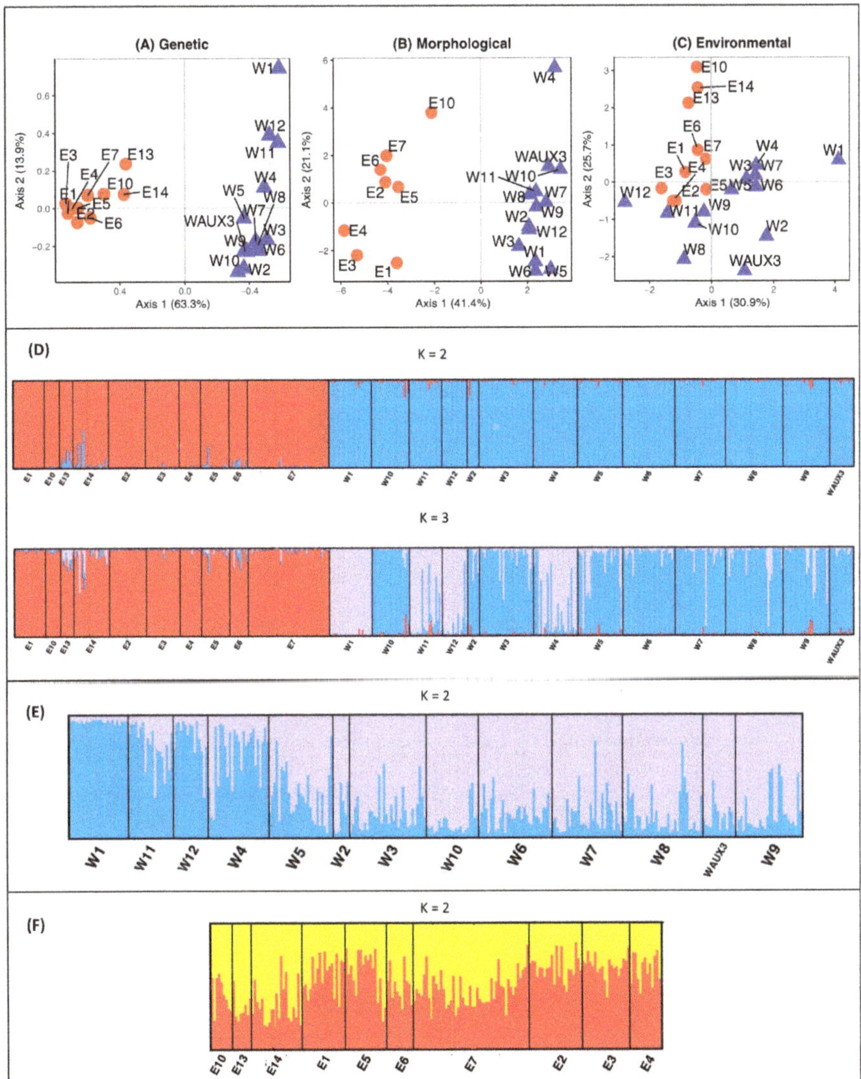

Figure 3. Principal coordinate analysis (PCoA) and principal component analyses (PCA) from 23 sampled populations from western (blue triangles) and eastern (red circles) populations of *Quercus havardii*. The percent of variation explained by each principal component is included on each axis. (**A**) PCoA of genetic variation among sampled populations shown by scores on the first two coordinates of variation at 11 microsatellite loci in 489 genotyped individuals; (**B**) PCA plot of mean population-level morphological variation on the first two coordinates derived from 12 characters in 928 measured leaves, consisting of 2 leaves per individual; (**C**) PCA plot of environmental variation on the first two components using 13 uncorrelated bioclimatic variables; (**D**) population genetic structure plots by STRUCTURE for range-wide with K = 2; (**E**) K = 3, within-region for eastern populations using optimal clustering value, K = 2; (**F**) within-region for western populations showing optimal clustering value, K = 2.

3.3. Environmental Differentiation, Variation, and Influence on Genetic Diversity

Principal component analysis of the six uncorrelated bioclimatic variables showed significant separation between populations from the East and West regions, with the first two components representing 56.6% of the variation (Figure 3C). More specifically, PC1 represented 30.9% of variation while PC2 represented 25.7% (Table A9). Maximum temperature of warmest month (BIO5), mean temperature of driest quarter (BIO9), and precipitation of coldest quarter (BIO19) showed almost equal contributions to the separation of populations along PC1, while annual temperature range (BIO7) and precipitation of the driest month (BIO14) showed along PC2 (Table A8). Mean temperature of the driest quarter was most important for the separation of West populations, while most East populations were primarily separated by mean diurnal range and precipitation of the driest and coldest months. We found higher average temperatures and precipitation in the East when compared to western populations (Table A7). Additionally, there is less seasonal temperature variability in the eastern region. Notably, one western population, W1, was found to be more climatically similar to populations in the East.

We found little evidence of environmental niche overlap (D = 0.002; Figure A7) between the eastern and western regions of *Q. havardii*. Additionally, western populations were observed to occupy a larger climatic niche space than eastern populations. Bioclimatic variables representing environmental extremes (e.g., precipitation of the driest month, mean temperature of the warmest quarter, and maximum temperature of the warmest month) contributed most to the separation of the East and West regions in environmental space. We found no significant differences for either niche equivalency or similarity between the two regions.

For each of the regression models with the lowest BIC, using a Benjamini–Hochberg procedure was found to be significant ($p < 0.05$) between genetic statistics and environmental variables, with the exception of the model of relatedness within the eastern region (Table 3). When examining the significant regressions across the overall geographic space (i.e., analyzing the East and West populations together), we found that the selected models showed a strong relationship mostly between temperature-related variables and all genetic diversity metrics especially with bioclimatic variables representing extreme or limiting environmental factors. For the range-wide model, allelic richness and relatedness were mostly found to be associated with seasonality-based variables, while heterozygosity was associated with variables pertaining to environmental extremes. Within the East, we found that heterozygosity and relatedness were primarily correlated with temperature extremes (mean temperature of the driest quarter; BIO9), while allelic richness was primarily correlated with precipitation extremes (precipitation of the warmest quarter; BIO18). Within the West, we found very strong correlations between both heterozygosity and allelic richness with environmental variables (both with adjusted $R^2 > 0.93$). More specifically, heterozygosity and relatedness were associated with extreme or annual trends of precipitation (precipitation of the driest month and annual precipitation; BIO14 and BIO12, respectively), while allelic richness was associated with temperature extremes (mean temperature of wettest quarter; BIO8).

3.4. Approximate Bayesian Computation (ABC) Estimate of Divergence Time

The support for the IM model over the I model is 0.758/0.232, equaling a Bayes Factor of 3.26, which is "moderate" but not "very strong" evidence [80,81]. The out-of-bag prior error rate is 42%, suggesting that it is difficult to distinguish between the two models. After untransformed values (Table A10) were converted to usable parameters (Table A11), the mean divergence time using the I model is 12,380 years, with confidence intervals ranging from 567 to 59,500 years. This suggests that divergence possibly occurred on a timescale of tens of thousands of years. We also present posterior parameter estimates based on all simulations run over both models (Table A11). When both models are included, the median estimate of the divergence time is 31,000 years ago, though the confidence intervals run from a few thousand to more than ten million. This wider confidence range seems logical because the divergence time estimate will vary with the migration rate. Additionally, the estimated population size in the West was consistently larger than in the East.

Table 3. Genetic–environmental correlations for each of the best-supported multiple linear regression (MLR) models for *Quercus havardii* at regional and local geographic levels. The *p*-values for the variable coefficients are shown, in addition to the adjusted R^2 and *p*-values for each reported model.

	Range-Wide MLR			East MLR			West MLR		
	He	*Ar*	*Rel*	*He*	*Ar*	*Rel*	*He*	*Ar*	*Rel*
BIO2: Mean Diurnal Range	0.0473	0.0013	0.0105	0.0049	0.0016	0.0194			
BIO3: Isothermality							0.0118		
BIO5: Max Temp of Warmest Month	0.1034								
BIO7: Annual Temp Range	0.0002	0.0019	0.0153					0.0008	
BIO8: Mean Temp of Wettest Quarter							0.0004	0.0501	0.0011
BIO9: Mean Temp of Driest Quarter	0.0181			0.1306	0.0451	0.0775			
BIO10: Mean Temp of Warmest Quarter								0.0023	
BIO12: Annual Precipitation							0.0001		0.0635
BIO14: Precipitation of Driest Month	0.0041						0.1390	0.0006	
BIO15: Precipitation Seasonality							0.0231	0.0002	
BIO18: Precipitation of Warmest Quarter					0.1484				
BIO19: Precipitation of Coldest Quarter								0.0000	
Model Adjusted R^2	0.6121	0.5019	0.3457	0.6286	0.7527	0.5093	0.9303	0.9665	0.7257
Model *p*-value	0.0026	0.0007	0.0111	0.0259	0.0275	0.0687	0.0005	0.0003	0.0012

He: expected heterozygosity; *Ar*: allelic richness by locus; *Rel*: relatedness within a population.

4. Discussion

4.1. General Patterns of Genetic, Morphological, and Environmental Variation

We observed strong differentiation of genetic, environmental, and morphological datasets coinciding with contemporary geographic separation of populations. Across its entire geographic range, we found that *Q. havardii* has relatively moderate levels of genetic diversity and high differentiation when compared to other oaks [41,42,82,83]. While morphological differentiation, particularly population means, showed separation between the two regions principally driven by differences in leaf length, the total variance in the dataset was shown to be very high within populations. This suggests that morphological diversity and plasticity can be quite great at the individual and population level, which is a pattern typically seen and documented in oaks of different groups attributed to patterns of interspecific hybridization [84] or canopy position and light exposure [85].

Climate is likely playing a role in shaping genetic diversity within and among populations and regions of *Q. havardii*. We found stark differences in regional climate between the East and West regions (e.g., mean annual precipitation in eastern populations was nearly double that of western populations). Overall, in western populations, allelic richness is 25% higher and heterozygosity is 11% higher, while relatedness is <50%, compared to eastern populations. We also observed several correlations within regions of population genetic diversity statistics and environmental variables that have potential biological relevance to population size and persistence. Specifically, measures of genetic diversity (i.e., heterozygosity, allelic richness, and relatedness) were correlated with climatic variables that represent seasonality or extremes (e.g., heterozygosity and allelic richness were correlated with mostly temperature seasonality or extremes in the East; while heterozygosity, allelic richness, and relatedness were correlated with both precipitation and temperature extremes in the West). This suggests that these factors may be influencing long-term population size and stability and thus genetic diversity via genetic drift. Studies in other desert plants [86] and oak species [87,88] correlating neutral microsatellite markers and climatic variables showed a similar pattern of differences at a regional level. Additional studies within oaks similarly noted seasonality-related variables as possible key factors shaping genetic diversity [89,90] and the increased rates of evolution for leaf habit [22]. Of course, these are correlations; further research on the effects of climate on oak diversity will be needed to elucidate the importance of these mechanisms. Possible adaptive trait differentiation corresponding to morphological, genetic and environmental differentiation patterns should be explored further.

An alternative influence on genetic diversity could be the relatively higher rates of human disturbance in the East. The Llano Estacado region of Texas and New Mexico has high rates of ranching and oil and gas development [30]. In such disturbed areas, population extinction and colonization may be higher, which could cause lower genetic diversity, higher relatedness, and higher genetic differentiation due to bottleneck effects. Additionally, we observed high rates of clonality, but much lower rates than were suggested by earlier allozyme data for *Q. havardii* [91]. Future work should investigate in more detail the spatial distribution of clones within populations, which might have profound effects on genetic differentiation, genotypic diversity and perhaps the long-term viability of the species. Thus, we cannot fully untangle the influences of genetic diversity in this system, but our results do present specific hypotheses to be investigated in more detail.

We found that the divergence of eastern and western populations most likely occurred on the scale of tens of thousands of years ago, according to our results from the ABC analysis. Although confidence intervals are broad, this divergence estimate implies that these populations have been isolated for hundreds of generations and are likely sister species or subspecies; however, this assumes an evolutionary relationship between these populations that cannot be confirmed by our work. Our estimates align with a broader study of the American oak clade, suggesting that *Q. havardii* likely arose in the early- to mid-Miocene, and also suggested a significant adaptive transition in ecological space, perhaps due to niche specialization [22]. The history of these populations was likely influenced by glacial cycles and consequent effects such as drying of the deserts of the Southwest and loss of suitable habitat. To resolve questions of evolutionary history, possible hybrid origins, and phylogeography, we are currently working to incorporate samples from both geographic regions into the phylogeny of North American oaks [92].

4.2. Possible Implications for Conservation and Taxonomy

Quercus havardii is an important keystone species that defines the sand shinnery ecosystem and provides a habitat to vulnerable species such as the lesser prairie chicken and dunes sagebrush lizard. The conservation status is currently listed as "Endangered" by the IUCN Red List, which notes that populations are facing increased threats of habitat loss and fragmentation arising from development for oil and gas industries and agriculture [30]. We observed relatively moderate genetic diversity and a low degree of clonality in this study suggesting that the species, and most of its populations, are not in imminent danger of genetic problems like inbreeding. However, in our study, the population W1, a putative lagging-edge population [93], was identified as extremely vulnerable due to high clonality, high inbreeding and low genetic diversity, and thus should be considered as a possible priority for conservation. Other isolated or lagging-edge populations should be investigated as well.

Additionally, the taxonomic uncertainty of *Q. havardii* remains as it is currently treated as a singular species over the full extent of its range. However, this has been debated over the years as some authors have called the western region its own taxon, *Q. havardii* var. *tuckerii* Welsh or *Q. welshii* R.A. Denham, and has also been speculated as a putative introgressed hybrid with *Q. gambelli* [31,94,95]. Our work provided evidence that both eastern and western entities are genetically and morphologically distinct, and occupy unique environmental niche space with little to no overlap. However, future phylogenetic work may still be needed to elucidate the evolutionary relationships of these taxa and to support any taxonomic changes. If taxonomic changes are deemed valid, a reassessment of the conservation status of *Q. havardii* and *Q. welshii* will be needed to incorporate regional-specific information such as population genetic data and modified area of occurrences.

4.3. Caveats

There are several important caveats that should be mentioned in this work. It must be noted that sampling was not systematic within and across each region, nor designed a priori to evenly cover different environmental variables but rather focused on the known locations of populations. It is possible that the relationships observed in this study could be artifacts of sampling or reflect a relationship with another unknown underlying variable. Moreover, we do not incorporate data on some variables that may be most important in determining a species demographic and genetic stability, (e.g., soil attributes). Additionally, the ABC analysis has extremely wide confidence intervals and assumes very simple models. An alternative hypothesis that these data do not support, but should not yet be ruled out yet, is that the divergence is deeper in time and that these are not sister taxa but rather have evolved similar morphology, habitat and habitat preference via convergent evolution.

Of additional importance, hybridization is possible among many oak taxa, and hybrids have been observed in many studies. Although oaks have been deemed taxonomically difficult due to high rates of intraspecific gene exchange, genetic investigations show that contemporary hybridization is usually observed at relatively low levels, such that species may maintain coherence [96]. We do not examine hybridization in this study, and only focus on individuals morphologically and ecologically consistent with *Q. havardii*; however, morphologies ascribed to *Q. eastwoodiae* or the *Q. x undulata* complex were observed (FLD and MOR herbarium specimens McCauley 700, 701, and 702). Additionally, several collected populations (W1, W4, W11, and W12) did not occur in deep, shifting sand and were noted as having other *Quercus* species located only a few kilometers away. It is likely that the highly specific and unique sand habitat of most populations of *Q. havardii* could be playing a role in observed diversity patterns by significantly limiting the opportunity for interspecific gene flow. Nonetheless, hybridization and its possible influence on diversity and population persistence is an important future area to investigate for this species.

5. Conclusions

We have provided the first range-wide genetic study of one of the arid-adapted oaks of the Southwest USA and one of the relatively few integrated genetic, ecological, and morphological characterization of oaks in general. *Q. havardii* has somewhat lower genetic diversity and higher genetic differentiation when compared with other widespread and important oaks [97], as might be expected with this species' highly specialized and fragmented habitat. We find strong differentiation between eastern and western regions, potentially at the level of evolutionary significant units (ESUs) or higher, suggesting important conservation designation. Within this larger regional context, correlations with environmental variables suggest that temperature seasonality, precipitation extremes, and other key factors may be influencing local levels of genetic diversity. Future studies of arid-adapted oaks are needed as climates continue to rapidly change and make habitats increasingly vulnerable, thus emphasizing the importance of these species as possible sources of genetic resources for breeding or restoration.

Author Contributions: Conceptualization, S.H., D.D. and R.A.M.; methodology, B.A.Z., D.D., E.S.; software, B.A.Z., S.H., R.A.M.; validation, B.A.Z., S.H., R.A.M., D.D., E.S.; formal analysis, B.A.Z., S.H., R.A.M., D.D.; investigation, B.A.Z., S.H., R.A.M., I.J.F., D.D., E.S.; resources, S.H. and R.A.M.; data curation, B.A.Z., R.A.M., S.H.; writing—original draft preparation, B.A.Z. and S.H.; writing— review and editing, B.A.Z., S.H., R.A.M., I.J.F., D.D., E.S.; visualization, B.A.Z., R.A.M., E.S., S.H.;

supervision, S.H.; project administration, S.H.; funding acquisition, S.H. All authors have read and agreed to the published version of the manuscript.

Funding: This research was funded by the U.S. Forest Service and American Public Gardens Association Tree Gene Conservation Partnership and also in part by The Morton Arboretum.

Data Availability Statement: The data and code for analyses used in this study are openly available on the Github repository Quercus_havardii_Forests_Manuscript at doi:10.5281/zenodo.4453098.

Acknowledgments: We would like to thank the three anonymous reviewers for their insightful comments on the manuscript. We thank Mackenzie Codon and Cindy Johnson for their assistance with lab work, and Kevin Feldheim and Isabel Distefano for providing their technical support on fragment analysis equipment and protocols. Additionally, we thank the Field Museum for providing us access to equipment for microsatellite analysis. The authors would like to thank the following permit-granting agencies: Bureau of Land Management, Texas Parks and Wildlife Department, New Mexico Department of Fish and Game, The Nature Conservancy, US Forest Service, Navajo Nation Department of Fish and Wildlife. We also thank Robert D. Cox, Jamie Baker, Brandon Childers, and Mike Melendrez for their assistance with the collection of material. The authors also thank Drs. Jamie Gillooly and Al Meyers for their valuable comments that greatly improved the manuscript.

Conflicts of Interest: The authors declare no conflict of interest. The funders had no role in the design of the study; in the collection, analyses, or interpretation of data; in the writing of the manuscript, or in the decision to publish the results.

Appendix A

Table A1. Microsatellite markers used for *Quercus havardii* population genetics.

Locus	Source [Citation]	Fluorescent Dye	Repeat Unit
QS03797	Chatwin et al. 2014 [98]	6-FAM	$(CA)_7$
MSQ13	Dow et al. 1995 [99]	6-FAM	$(TC)_{14}$
QrZAG20	Kampfer et al. 2004 [100]	VIC	$(TC)_{18}$
QpZAG110	Steinkellner et al. 1997 [101]	NED	$(AG)_{15}$
QS00314	Chatwin et al. 2014 [98]	PET	$(GAA)_6$
QS1904	Chatwin et al. 2014 [98]	6-FAM	$(TC)_{10}$
MSQ4	Dow et al. 1995 [99]	VIC	$(AG)_{17}$
QS00562	Steinkellner et al. 1997 [101]	6-FAM	$(GA)_7$
QrZAG87	Kampfer et al. 2004 [100]	NED	$(TC)_{20}$
QpZAG1.5	Steinkellner et al. 1997 [101]	PET	$(GT)_5(GA)_9$
QM69.2M1	Isagi and Suhandono 1997 [102]	PET	$(TGG)_6(CGG)(TGG)_2$

Table A2. Genetic summary statistics by region.

Parameter	East Mean	West Mean	T-test *p*-Value	Wilcoxon *p*-Value
Allelic richness	47.160	58.877	0.0003	0.0013
Heterozygosity	0.587	0.653	0.0150	0.0148
Pairwise F_{ST}	0.085	0.079	0.3263	0.1111
Relatedness	0.176	0.077	0.0063	0.0093

Table A3. Eigenvalues and percent variability for each principal component for the PCA using 17 uncorrelated morphological variables presented in Figure 3B.

	Eigenvalue	Variance (%)	Cumulative Variance (%)
PC1	7.039	41.4	41.4
PC2	3.584	21.1	62.5
PC3	3.290	19.4	81.8
PC4	1.459	8.6	90.4
PC5	0.679	4.0	94.4
PC6	0.494	2.9	97.3
PC7	0.178	1.0	98.4
PC8	0.093	0.5	98.9
PC9	0.075	0.4	99.4
PC10	0.045	0.3	99.6
PC11	0.032	0.2	99.8

Table A3. *Cont.*

	Eigenvalue	Variance (%)	Cumulative Variance (%)
PC12	0.021	0.1	99.9
PC13	0.007	0.0	100.0
PC14	0.003	0.0	100.0
PC15	0.001	0.0	100.0
PC16	0.000	0.0	100.0
PC17	0.000	0.0	100.0

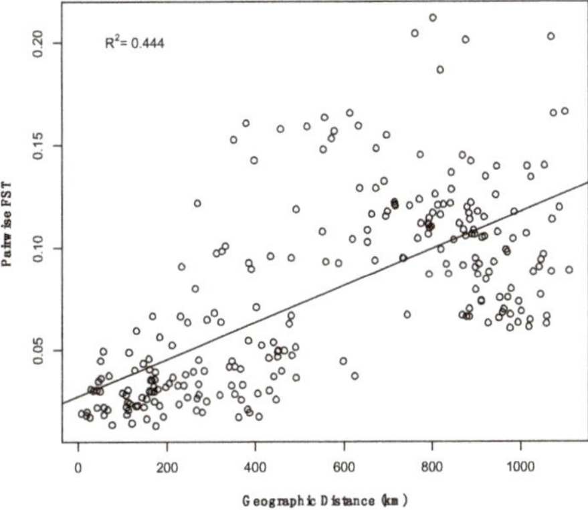

Figure A1. Mantel test to evaluate isolation by distance within *Quercus havardii*. Pairwise F_{ST} is compared to geographic distance (km) between each pair of populations.

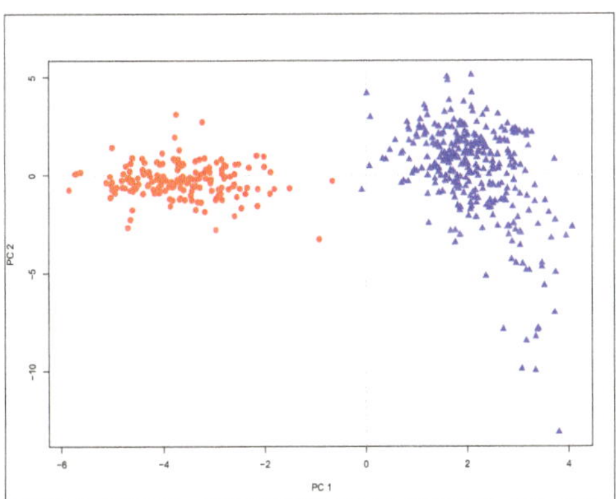

Figure A2. PCoA comparing genetic characteristics among 489 individuals of *Quercus havardii*. Western and eastern individuals are represented by blue triangles and red circles, respectively.

Table A4. Plot loading scores of principle components axes with eigenvalues greater than 1 for the PCA using 17 uncorrelated morphological variables presented in Figure 3B. Important loadings that contribute more than one variable's worth of information are in bold.

Character	PC1	PC2	PC3	PC4
Lobes	**0.32**	0.07	−0.20	0.19
Length	−0.35	0.08	−0.18	−0.08
BLBW	−0.26	−0.21	−0.25	0.09
MLBW	0.08	−0.15	−0.51	−0.12

Table A4. *Cont.*

Character	PC1	PC2	PC3	PC4
ALBW	−0.22	−0.14	−0.02	**−0.59**
LVL	**−0.31**	0.18	−0.19	−0.04
UVL	−0.04	0.19	**−0.49**	−0.05
UMLS	−0.08	**−0.43**	−0.22	−0.19
LMLS	−0.20	**−0.37**	−0.20	0.12
LLA	**0.35**	−0.07	−0.10	−0.15
ULA	−0.19	**0.31**	0.16	−0.17
MLA	**0.26**	0.09	−0.06	−0.41
LWR_1	0.20	−0.36	0.01	**0.25**
LWR_2	**0.35**	−0.17	−0.11	0.03
LWR_3	0.20	−0.23	**0.25**	−0.42
LODR_1	0.17	**0.33**	−0.28	0.11
LODR_2	0.24	**0.29**	−0.22	−0.24

The bolding here highlights the characters that most influence PCA's.

Table A5. Correlation of variables with the first two axes in a discriminant analysis of morphological variation across the full range of *Q. havardii*. The three most influential characters for each axis are indicated in bold.

Character	Abbreviation	CV1	CV2
Lobe Number	LOBES	**−0.332**	−0.068
Leaf Length	LENGTH	0.052	0.025
Basal Lobe Blade Width	BLBW	0.018	**−0.300**
Middle Lobe Blade Width	MLBW	−0.135	**0.299**
Apical Lobe Blade Width	ALBW	**0.058**	**−0.392**
Lower Vein Length	LVL	0.041	−0.057
Upper Vein Length	UVL	0.018	−0.084
Upper Middle Lobe Sinus Width	UMLS	0.008	0.174
Lower Middle Lobe Sinus Width	LMLS	**0.065**	−0.075
Lower Lobe Angle	LLA	−0.034	0.007
Upper Lobe Angle	ULA	0.014	0.013
Middle Lobe Angle	MLA	0.001	−0.013
Length to Width Ratio 1 (BLBW/Length)	LWR_1	**−0.332**	−0.068
Length to Width Ratio 2 (MLBW/Length)	LWR_2	0.052	0.025
Length to Width Ratio 3 (ALBW/Length)	LWR_3	0.018	**−0.300**
Lobe Depth Ratio 1	LODR_1	−0.135	**0.299**
Lobe Depth Ratio 2	LODR_2	**0.058**	**−0.392**

The bolding here highlights the characters that most influence PCA's.

Table A6. Uncorrelated bioclimatic variables used as inputs for the principle component analysis (Figure 1), building multiple linear regressions (MLR) models for each of the three regional and local geographic level datasets (Table 3), and variables used for the environmental niche overlap analysis (Figure A7).

Bioclimatic Variable	Code	Range-Wide MLR and PCA	East MLR	West MLR	Niche Overlap
Annual Mean Temperature	BIO1				
Mean Diurnal Range	BIO2	X	X		
Isothermality	BIO3			X	
Temperature Seasonality	BIO4		X		
Maximum Temperature of Warmest Month	BIO5	X			X
Minimum Temperature of Coldest Month	BIO6				
Temperature Annual Range	BIO7	X		X	X
Mean Temperature of Wettest Quarter	BIO8		X	X	X
Mean Temperature of Driest Quarter	BIO9	X	X		X
Mean Temperature of Warmest Quarter	BIO10			X	
Mean Temperature of Coldest Quarter	BIO11				
Annual Precipitation	BIO12			X	
Precipitation of Wettest Month	BIO13				
Precipitation of Driest Month	BIO14	X		X	X
Precipitation Seasonality	BIO15			X	X
Precipitation of Wettest Quarter	BIO16				
Precipitation of Driest Quarter	BIO17				
Precipitation of Warmest Quarter	BIO18		X		X
Precipitation of Coldest Quarter	BIO19	X		X	

Table A7. Summary of the bioclimatic variable means from the 23 populations of *Quercus havardii* sampled in this study.

Pop	BIO1	BIO2	BIO3	BIO4	BIO5	BIO6	BIO7	BIO8	BIO9	BIO10	BIO11	BIO12	BIO13	BIO14	BIO15	BIO16	BIO17	BIO18	BIO19
E1	15	17.1	44	801.7	34.1	−4.8	38.9	23.8	5	24.7	5	463	69	12	61.6	201	42	198	42
E2	15.1	17.2	43.8	801.9	34.3	−4.9	39.2	24.1	5.1	24.8	5.1	385	70	8	68.8	190	31	179	31
E3	14.2	17.7	44.8	797.2	33.5	−5.9	39.4	23.8	5.9	23.8	4.3	412	74	10	69.9	199	32	199	32
E4	14.3	17.9	45.1	798.2	33.8	−5.9	39.7	23.9	4.3	23.9	4.3	421	74	9	69	200	32	200	32
E5	14.5	15	36.8	929.4	34.7	−5.9	40.7	23.5	2.9	25.9	2.9	528	89	9	58.1	219	42	195	42
E6	14.6	14.8	37	906	34.6	−5.4	40	23.4	3.3	25.7	3.3	568	93	13	56	228	49	198	49
E7	14.5	14.5	35.9	927.2	34.6	−5.9	40.5	18.9	2.9	25.8	2.9	570	96	12	54.8	229	51	193	51
E10	17.5	14.8	39.7	808.5	35.6	−1.6	37.2	26.5	7.1	27	7.1	492	87	16	55.6	184	53	157	53
E13	15.9	14.7	39.5	820.7	34.1	−3	37.1	24.1	5.6	25.8	5.6	541	78	14	55.1	208	52	203	52
E15	16.4	15.1	39.6	837	35.3	−2.7	38.1	24.7	5.9	26.6	5.9	543	79	17	53	201	57	196	57
W1	17.1	19.7	47.6	793.9	37.8	−3.5	41.3	26.7	20.2	27.1	7.7	377	53	7	41.3	137	38	103	99
W2	12.2	17.1	38.9	956	33.1	−10.9	44	19.1	16.5	24.1	0.5	214	27	6	29.4	72	32	51	51
W3	11.8	15.9	38.7	906.2	32.1	−9.1	41.1	22.4	15.8	23.3	0.9	233	31	7	34.3	82	32	65	53
W4	12.3	16.4	40.4	876.8	32.4	−8.2	40.6	22.7	16.1	23.5	1.9	262	35	8	32.4	84	35	68	69
W5	12.1	16	39.1	903.7	32.4	−8.6	41	22.7	16.2	23.5	1.3	229	32	7	33.1	79	32	64	54
W6	12.3	14.2	35.1	943.8	32.1	−8.3	40.3	19	16.5	24.2	1	208	26	7	32.5	73	30	55	45
W7	11.9	13.2	34	926.1	30.8	−8	38.7	18.6	15.9	23.6	0.8	233	28	8	29.8	79	34	60	56
W8	12.2	16	36.3	996.7	33.2	−10.7	43.9	19	−0.1	24.6	−0.1	179	22	7	31.3	64	33	47	33
W9	12.6	16.3	36.5	1002.6	33.7	−10.8	44.5	19.6	22.3	25.1	0.2	245	31	9	24.9	74	51	53	52
W10	12	17	36.3	1041.7	33.8	−13.1	46.9	19.2	−1.1	24.8	−1.1	226	29	9	25.3	71	47	50	47
W11	12.5	19.4	45.6	847.6	33.4	−9.1	42.5	22.3	7.7	23.1	2.2	239	48	10	62.9	120	32	98	35
W12	9.1	15	41.7	784.5	27.4	−8.6	35.9	19.3	0	19.3	0	364	65	9	58.2	161	33	161	33
WAUX3	12.3	16.2	36.5	1003.8	33.6	−10.8	44.4	19.1	0	24.8	0	177	21	7	29.9	62	33	46	33
East Mean	15.2	15.9	40.6	842.8	34.5	−4.6	39.1	23.7	4.8	25.4	4.6	492.3	80.9	12.0	60.2	205.9	44.1	191.8	44.1
East ± SD	1.1	1.4	3.5	55.5	0.7	1.6	1.3	1.9	1.4	1.1	1.5	68.2	9.7	3.1	6.7	15.1	9.7	13.9	9.7
West Mean	12.3	16.3	39.0	921.8	32.8	−9.2	41.9	20.7	11.2	23.9	1.2	245.1	34.5	7.8	35.8	89.1	35.5	70.8	50.8
West ± SD	1.7	1.8	4	80.8	2.3	2.3	2.9	2.4	8.6	1.8	2.2	60.6	13	1.2	11.8	30.5	6.3	32.5	18.1

Table A8. Plot loading scores of principle components axes with eigenvalues greater than 1 for the PCA using bioclimatic data from 6 uncorrelated variables presented in Figure 3C. Important loadings that contribute more than one variable's worth of information are in bold.

Bioclimatic Variable	PC1	PC2	PC3
BIO2: Mean Diurnal Range	−0.113	−0.268	0.331
BIO5: Max Temp of Warmest Month	**0.410**	0.062	**0.663**
BIO7: Annual Temp Range	0.101	−0.635	0.375
BIO9: Mean Temp of Driest Quarter	**0.623**	−0.002	−0.389
BIO14: Precipitation of Driest Month	−0.334	**0.610**	0.364
BIO19: Precipitation of Coldest Quarter	**0.556**	0.385	0.164

The bolding here highlights the characters that most influence PCA's.

Table A9. Eigenvalues and percent variability for each principal component for the PCA using bioclimatic data from 6 uncorrelated variables presented in Figure 3C.

	Eigenvalue	Variance (%)	Cumulative Variance (%)
PC1	1.852	30.9	30.9
PC2	1.542	25.7	56.6
PC3	1.322	22.0	78.6
PC4	0.921	15.3	94.0
PC5	0.275	4.6	98.5
PC6	0.088	1.5	100.0

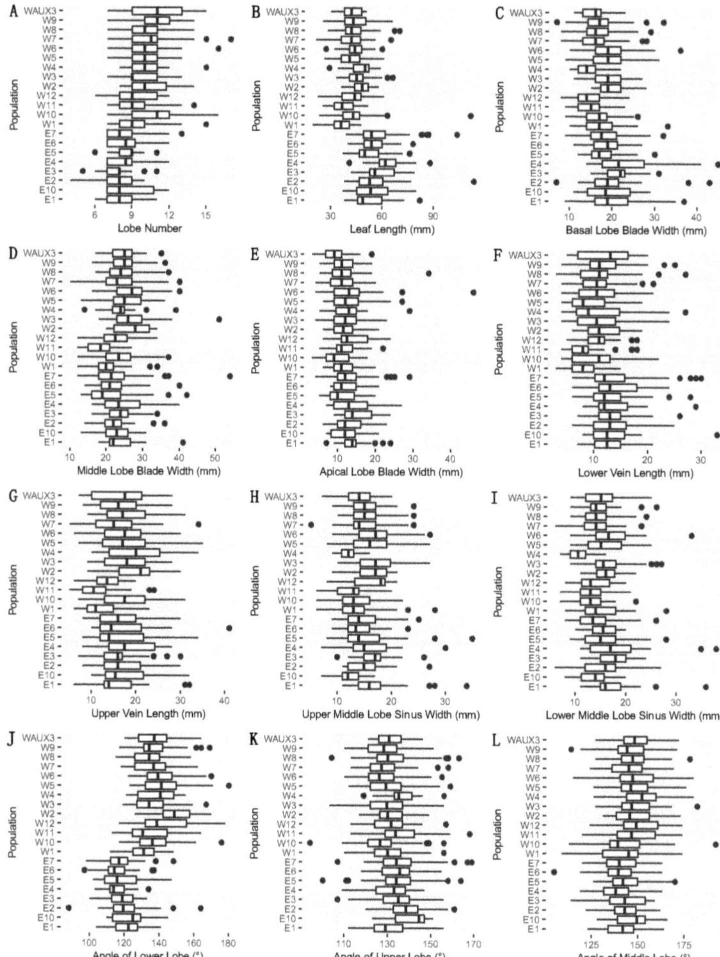

Figure A3. Box plots of 12 directly measured variables in 21 populations of *Quercus havardii* showing (**A**) leaf lobe number; (**B**) leaf length; (**C**) basal lobe blade width; (**D**) middle lobe blade width; (**E**) apical lobe blade width (**F**) lower vein length; (**G**) upper vein length; (**H**) upper middle lobe sinus width; (**I**) lower middle lobe sinus width; (**J**) angle of lower lobe; (**K**) angle of upper lobe; and (**L**) angle of middle lobe. Boxes depict standard Tukey representations. Population names follow Table 1.

Table A10. Untransformed parameter estimates from approximate Bayesian computation analysis for splitting time (TS), migration (M), and theta 1 and theta 2 (representing the East and West regions, respectively) under the isolation with migration (IM) and isolation only (I) models.

Model	Parameter	0.025 CI	Median	0.975 CI
IM	Theta 1	0.396	2.17	16.9
IM	Theta 2	2.35	4.1	75.5
IM	M	1.75×10^{-10}	0.177	5.52
IM	TS	0.0157	0.181	76.7
I	Theta 1	0.1816	2.105	40.96
I	Theta 2	1.938	3.47	10.43
I	M	1.75×10^{-10}	0.1766	5.518
I	TS	0.003393	0.07413	0.3564

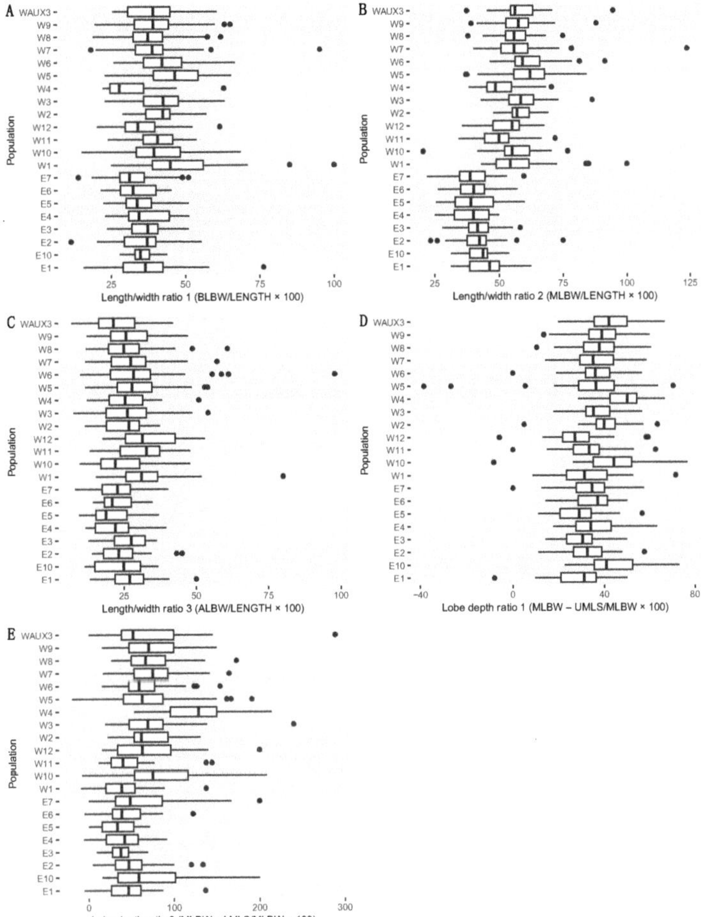

Figure A4. Box plots of five ratio variables derived from directly measured variables in 21 populations of *Quercus havardii* showing (**A**) length/width ratio 1 (BLBW/LENGTH); (**B**) length/width ratio 2 (MLBW/LENGTH); (**C**) length/width ratio 3 (ALBW/LENGTH); and (**D**) lobe depth ratio (MLBW—UMLS/MLBW); (**E**) lobe depth ratio 2 (MLBW—LMLS/MLBW). Boxes depict standard Tukey representations. Population names follow Table 1.

Table A11. Transformed parameter estimates from approximate Bayesian computation analysis for splitting time (TS), migration (M), and theta 1 and theta 2 (representing the East and West regions, respectively) under the isolation with migration (IM) and isolation only (I) models.

Model	Parameter	0.025 CI	Median	0.975 CI
IM	Theta 1	157	863	6710
IM	Theta 2	930	1630	29,900
IM	M	5.07×10^{-10}	5.12×10^{-5}	1.60×10^{-3}
IM	TS	2710	31,100	13,200,000
I	Theta 1	72	835	16,300
I	Theta 2	769	1377	4140
I	M	5.23×10^{-10}	5.29×10^{-5}	1.65×10^{-3}
I	TS	567	12,380	59,500

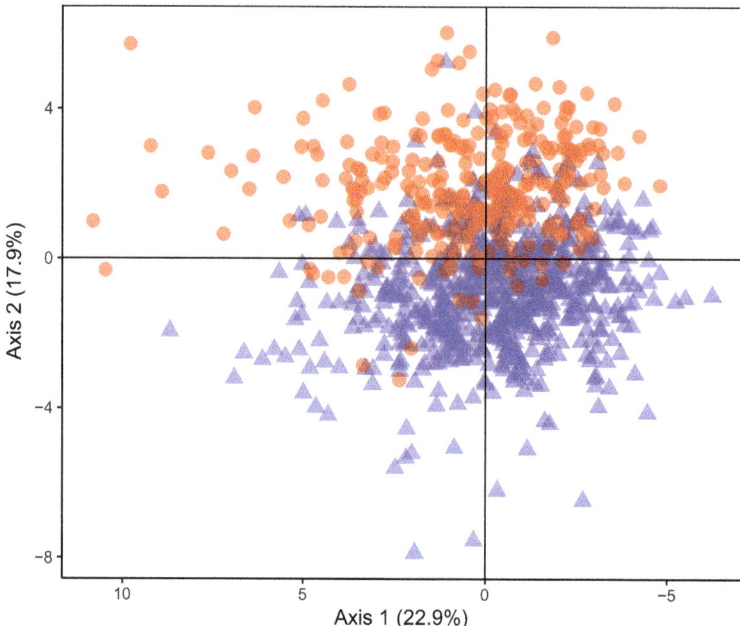

Figure A5. PCA comparing morphological characteristics among all sampled individuals of *Quercus havardii*, with the first two components explaining 40.8% of variation. Individuals from the western and eastern regions are represented by blue triangles and red circles, respectively.

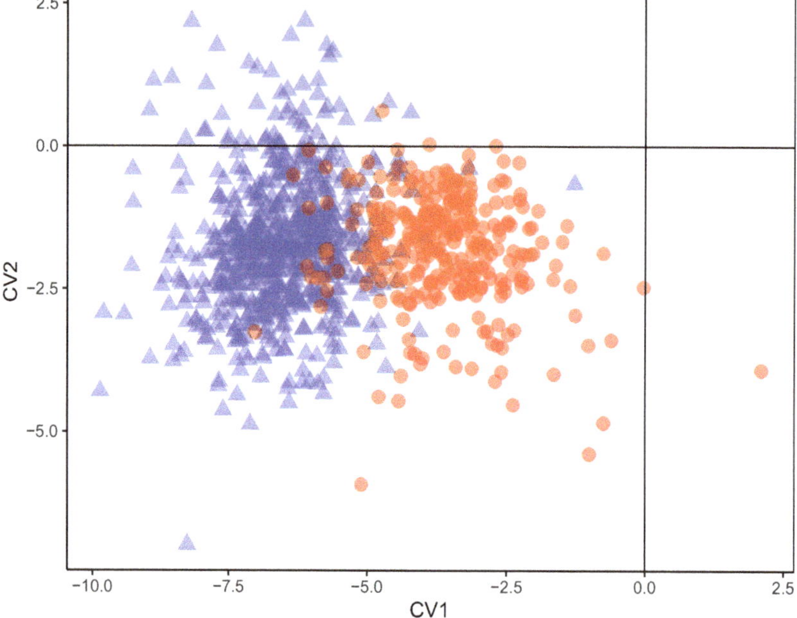

Figure A6. CVA plot of the first two canonical variates axes depicting segregation between western and eastern groups of *Q. harvardii*. Western and eastern individuals are represented by blue triangles and red circles, respectively.

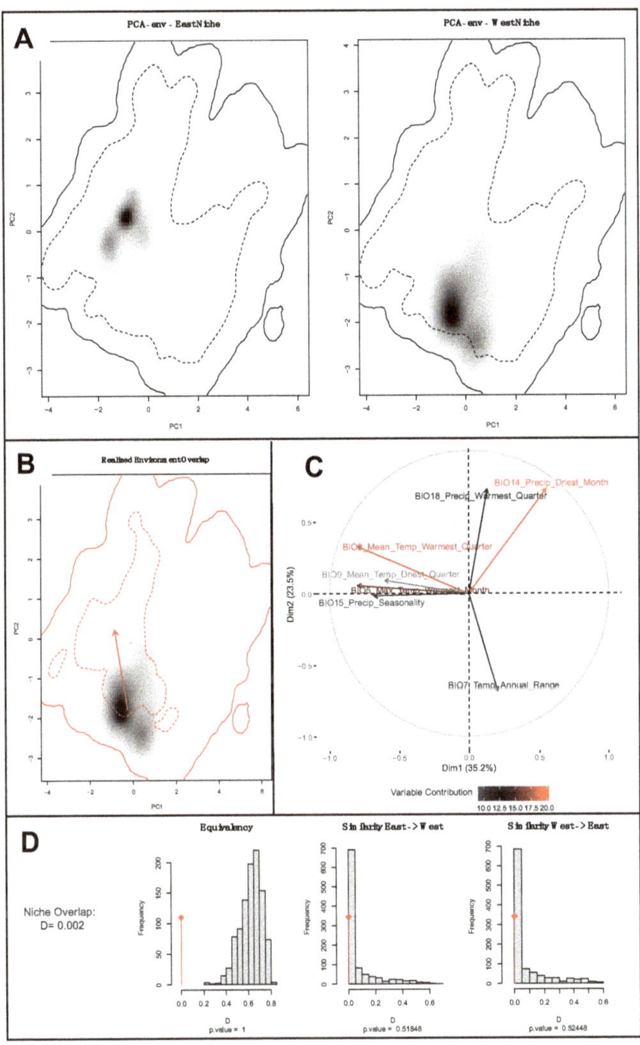

Figure A7. Environmental niche overlap analysis of *Quercus havardii* populations from the eastern and western regions of the species' disjunct range showing: (**A**) plot of the occupancy of the environment by populations located in the East (left) and West (right); (**B**) area of niche unique to eastern populations (red), western populations (blue), and estimated area of shared overlap (green); (**C**) correlation circle of the loadings of individual environmental variables to the two PCA axes; (**D**) estimated niche overlap values for the East and West regions in terms of Schoener's D, and niche similarity and equivalency tests with corresponding *p*-values.

References

1. Shafer, A.B.A.; Wolf, J.B.W. Widespread Evidence for Incipient Ecological Speciation: A Meta-Analysis of Isolation-by-Ecology. *Ecol. Lett.* **2013**, *16*, 940–950. [CrossRef] [PubMed]
2. Coyne, J.A. Speciation in a Small Space. *Proc. Natl. Acad. Sci. USA* **2011**, *108*, 12975–12976. [CrossRef] [PubMed]
3. Slatkin, M. Gene Flow and the Geographic Structure of Natural Populations. *Science* **1987**, *236*, 787–792. [CrossRef]
4. Ellstrand, N.C.; Elam, D.R. Population Genetic Consequences of Small Population Size: Implications for Plant Conservation. *Annu. Rev. Ecol. Syst.* **1993**, *24*, 217–242. [CrossRef]

5.   Jain, S.K.; Bradshaw, A.D. Evolutionary Divergence among Adjacent Plant Populations I. The Evidence and Its Theoretical Analysis. *Heredity* **1966**, *21*, 407–441. [CrossRef]
6.   Schluter, D. Ecology and the Origin of Species. *Trends Ecol. Evol.* **2001**, *16*, 372–380. [CrossRef]
7.   Rundle, H.D.; Nosil, P. Ecological Speciation. *Ecol. Lett.* **2005**, *8*, 336–352. [CrossRef]
8.   Thorne, R.F. Major Disjunctions in the Geographic Ranges of Seed Plants. *Q. Rev. Biol.* **1972**, *47*, 365–411. [CrossRef]
9.   Honnay, O.; Verheyen, K.; Butaye, J.; Jacquemyn, H.; Bossuyt, B.; Hermy, M. Possible Effects of Habitat Fragmentation and Climate Change on the Range of Forest Plant Species. *Ecol. Lett.* **2002**, *5*, 525–530. [CrossRef]
10.   Beatty, G.E.; Provan, J. Post-Glacial Dispersal, Rather than in Situ Glacial Survival, Best Explains the Disjunct Distribution of the Lusitanian Plant Species *Daboecia cantabrica* (Ericaceae). *J. Biogeogr.* **2012**, *40*, 335–344. [CrossRef]
11.   Hewitt, G.M. Post-Glacial Re-Colonization of European Biota. *Biol. J. Linn. Soc.* **1999**, *68*, 87–112. [CrossRef]
12.   Zhao, M.; Chang, Y.; Kimball, R.T.; Zhao, J.; Lei, F.; Qu, Y. Pleistocene Glaciation Explains the Disjunct Distribution of the Chestnut-Vented Nuthatch (Aves, Sittidae). *Zool. Scr.* **2018**, *48*, 33–45. [CrossRef]
13.   Grant, V. *Plant Speciation*, 2nd ed.; Columbia University Press: New York, NY, USA, 1981.
14.   Ortego, J.; Riordan, E.C.; Gugger, P.F.; Sork, V.L. Influence of Environmental Heterogeneity on Genetic Diversity and Structure in an Endemic Southern Californian Oak. *Mol. Ecol.* **2012**, *21*, 3210–3223. [CrossRef] [PubMed]
15.   Ülker, E.D.; Tavşanoğlu, Ç.; Perktaş, U. Ecological Niche Modelling of Pedunculate Oak (*Quercus robur*) Supports the 'Expansion–Contraction'Model of Pleistocene Biogeography. *Biol. J. Linn. Soc.* **2018**, *123*, 338–347. [CrossRef]
16.   Di Pietro, R.; Conte, A.L.; Di Marzio, P.; Fortini, P.; Farris, E.; Gianguzzi, L.; Müller, M.; Rosati, L.; Spampinato, G.; Gailing, O. Does the Genetic Diversity among Pubescent White Oaks in Southern Italy, Sicily and Sardinia Islands Support the Current Taxonomic Classification? *Eur. J. For. Res.* **2020**, *140*, 355–371. [CrossRef]
17.   Fortini, P.; Antonecchia, G.; Di Marzio, P.; Maiuro, L.; Viscosi, V. Role of Micromorphological Leaf Traits and Molecular Data in Taxonomy of Three Sympatric White Oak Species and Their Hybrids (*Quercus* L.). *Plant Biosyst. Int. J. Deal. All Asp. Plant Biol.* **2015**, *149*, 546–558.
18.   Fortini, P.; Di Marzio, P.; Di Pietro, R. Differentiation and Hybridization of *Quercus frainetto*, *Q. petraea*, and *Q. pubescens*, (Fagaceae): Insights from Macro-Morphological Leaf Traits and Molecular Data. *Plant Syst. Evol.* **2015**, *301*, 375–385. [CrossRef]
19.   Fortini, P.; Viscosi, V.; Maiuro, L.; Fineschi, S.; Vendramin, G. Comparative Leaf Surface Morphology and Molecular Data of Five Oaks of the Subgenus *Quercus* Oerst (Fagaceae). *Plant Biosyst.* **2009**, *143*, 543–554. [CrossRef]
20.   Manos, P.S.; Doyle, J.J.; Nixon, K.C. Phylogeny, Biogeography, and Processes of Molecular Differentiation in *Quercus* Subgenus *Quercus* (Fagaceae). *Mol. Phylogenetics Evol.* **1999**, *12*, 333–349. [CrossRef] [PubMed]
21.   Cavender-Bares, J.; González-Rodríguez, A.; Eaton, D.A.; Hipp, A.A.; Beulke, A.; Manos, P.S. Phylogeny and Biogeography of the American Live Oaks (*Quercus* Subsection *Virentes*): A Genomic and Population Genetics Approach. *Mol. Ecol.* **2015**, *24*, 3668–3687. [CrossRef]
22.   Hipp, A.L.; Manos, P.S.; González-Rodríguez, A.; Hahn, M.; Kaproth, M.; McVay, J.D.; Avalos, S.V.; Cavender-Bares, J. Sympatric Parallel Diversification of Major Oak Clades in the Americas and the Origins of Mexican Species Diversity. *New Phytol.* **2017**, *217*, 439–452. [CrossRef] [PubMed]
23.   Cavender-Bares, J. Diversification, Adaptation, and Community Assembly of the American Oaks (*Quercus*), a Model Clade for Integrating Ecology and Evolution. *New Phytol.* **2018**, *221*, 669–692. [CrossRef] [PubMed]
24.   Poulos, H.M. A Review of the Evidence for Pine-Oak Niche Differentiation in the American Southwest. *J. Sustain. For.* **2009**, *28*, 92–107. [CrossRef]
25.   Felger, R.S.; Johnson, M.B. Biodiversity and Management of the Madrean Archipelago: The Sky Islands of Southwestern United States and Northwestern Mexico. In *Biodiversity and Management of the Madrean Archipelago: The Sky Islands of Southwestern United States and northern Mexico*; The Station: Fort Collins, CO, USA, 1995; Volume 264, pp. 71–77.
26.   Dhillion, S.S.; Mills, M.H. The sand shinnery oak (*Quercus havardii*) communities of the Llano Estacado: History, structure, ecology, and restoration. In *Savannas, Barrens, and Rock Outcrop Plant Communities of North America*; Cambridge University Press: Cambridge, UK, 1999; pp. 262–274.
27.   Peterson, R.; Boyd, C.S. *Ecology and Management of Sand Shinnery Communities: A Literature Review*; U.S. Department of Agriculture, Forest Service, Rocky Mountain Research Station: Fort Collins, CO, USA, 2000.
28.   Dhillion, S.S. Environmental Heterogeneity, Animal Disturbances, Microsite Characteristics, and Seedling Establishment in a *Quercus havardii* Community. *Restor. Ecol.* **1999**, *7*, 399–406. [CrossRef]
29.   Davis, W.J. *Shin-Oak (Quercus havardii, Rydb.*; Fagaceae) Rhizome Shoot Production: Possibilities for Use in Restoration. M.S.; Texas Tech University: Lubbock, TX, USA, 2013.
30.   Jerome, D.; Beckman, E.; Kenny, L.; Wenzell, K.; Kua, C.; Westwood, M. *The Red List of US Oaks*; The Morton Arboretum: Lisle, IL, USA, 2017.
31.   Tucker, J.M. Studies in the *Quercus undulata* Complex. IV. The Contribution of *Quercus havardii*. *Am. J. Bot.* **1970**, *57*, 71–84. [CrossRef]
32.   McCauley, R.A.; Christie, B.J.; Ireland, E.L.; Landers, R.A.; Nichols, H.R.; Schendel, M.T. Influence of Relictual Species on the Morphology of a Hybridizing Oak Complex: An Analysis of the *Quercus* x *undulata* Complex in the Four Corners Region. *West. N. Am. Nat.* **2012**, *72*, 296–310. [CrossRef]

33. Jarčuška, B.; Milla, R. Shoot-Level Biomass Allocation Is Affected by Shoot Type in *Fagus sylvatica*. *J. Plant Ecol.* **2012**, *5*, 422–428. [CrossRef]

34. Cicák, A. Estimation of Morphological Parameters for European Beech Leaves on Spring Shoots Using the Method of Calculation Coefficients. *Oecol* **2003**, *30*, 131–140.

35. Viscosi, V.; Lepais, O.; Gerber, S.; Fortini, P. Leaf Morphological Analyses in Four European Oak Species (*Quercus*) and Their Hybrids: A Comparison of Traditional and Geometric Morphometric Methods. *Plant Biosyst.* **2009**, *143*, 564–574. [CrossRef]

36. Albarrán-Lara, A.L.; Mendoza-Cuenca, L.; Valencia-Avalos, S.; González-Rodríguez, A.; Oyama, K. Leaf Fluctuating Asymmetry Increases with Hybridization and Introgression between *Quercus magnoliifolia* and *Quercus resinosa* (Fagaceae) through an Altitudinal Gradient in Mexico. *Int. J. Plant Sci.* **2010**, *171*, 310–322. [CrossRef]

37. Holmgren, P.K.; Holmgren, N.H. Index Herbariorum. *Taxon* **1991**, *40*, 687–692. [CrossRef]

38. Holleley, C.E.; Geerts, P.G. Multiplex Manager 1.0: A Cross-Platform Computer Program That Plans and Optimizes Multiplex PCR. *BioTechniques* **2009**, *46*, 511–517. [CrossRef]

39. Dzialuk, A.; Chybicki, I.; Burczyk, J. PCR Multiplexing of Nuclear Microsatellite Loci in *Quercus* Species. *Plant Mol. Biol. Rep.* **2005**, *23*, 121–128. [CrossRef]

40. Abraham, S.T.; Zaya, D.N.; Koenig, W.D.; Ashley, M.V. Interspecific and Intraspecific Pollination Patterns of Valley Oak, *Quercus lobata*, in a Mixed Stand in Coastal Central California. *Int. J. Plant Sci.* **2011**, *172*, 691–699. [CrossRef]

41. Backs, J.R.; Terry, M.; Ashley, M.V. Using Genetic Analysis to Evaluate Hybridization as a Conservation Concern for the Threatened Species *Quercus hinckleyi* C.H. Muller (Fagaceae). *Int. J. Plant Sci.* **2016**, *177*, 121–131. [CrossRef]

42. Craft, K.J.; Ashley, M.V. Landscape Genetic Structure of Bur Oak (*Quercus macrocarpa*) Savannas in Illinois. *For. Ecol. Manag.* **2007**, *239*, 13–20. [CrossRef]

43. Oosterhout, C.V.; Hutchinson, W.F.; Wills, D.P.; Shipley, P. Micro-Checker: Software for Identifying and Correcting Genotyping Errors in Microsatellite Data. *Mol. Ecol. Notes* **2004**, *4*, 535–538. [CrossRef]

44. Chybicki, I.J.; Burczyk, J. Simultaneous Estimation of Null Alleles and Inbreeding Coefficients. *J. Hered.* **2008**, *100*, 106–113. [CrossRef] [PubMed]

45. R Core Team. *R: A Language and Environment for Statistical Computing*; R Foundation for Statistical Computing: Vienna, Austria, 2020.

46. Kamvar, Z.N.; Tabima, J.F.; Grünwald, N.J. Poppr: An R Package for Genetic Analysis of Populations with Clonal, Partially Clonal, and/or Sexual Reproduction. *PeerJ* **2014**, *2*, e281. [CrossRef]

47. Jombart, T. Adegenet. A R Package for the Multivariate Analysis of Genetic Markers. *Bioinformatics* **2008**, *24*, 1403–1405. [CrossRef]

48. Goudet, J. Hierfstat, a Package for R to Compute and Test Hierarchical F-Statistics. *Mol. Ecol. Notes* **2005**, *5*, 184–186. [CrossRef]

49. Kraemer, P.; Gerlach, G. Demerelate: Calculating Interindividual Relatedness for Kinship Analysis Based on Codominant Diploid Genetic Markers Using R. *Mol. Ecol. Resour.* **2017**, *17*, 1371–1377. [CrossRef] [PubMed]

50. Peakall, R.; Smouse, P.E. GenAlEx 6.5: Genetic Analysis in Excel. Population Genetic Software for Teaching and Research–an Update. *Bioinformatics* **2012**, *28*, 2537–2539. [CrossRef] [PubMed]

51. Pritchard, J.K.; Stephens, M.; Donnelly, P. Inference of Population Structure Using Multilocus Genotype Data. *Genetics* **2000**, *155*, 945–959. [PubMed]

52. Earl, D.A.; vonHoldt, B.M. STRUCTURE HARVESTER: A Website and Program for Visualizing STRUCTURE Output and Implementing the Evanno Method. *Conserv. Genet. Resour.* **2011**, *4*, 359–361. [CrossRef]

53. Rosenberg, N.A. Distruct: A Program for the Graphical Display of Population Structure. *Mol. Ecol. Notes* **2003**, *4*, 137–138. [CrossRef]

54. Jombart, T.; Devillard, S.; Balloux, F. Discriminant Analysis of Principal Components: A New Method for the Analysis of Genetically Structured Populations. *BMC Genet.* **2010**, *11*, 94. [CrossRef]

55. Evanno, G.; Regnaut, S.; Goudet, J. Detecting the Number of Clusters of Individuals Using the Software Structure: A Simulation Study. *Mol. Ecol.* **2005**, *14*, 2611–2620. [CrossRef]

56. Uslu, E.; Bakış, Y. Morphometric Analyses of the Leaf Variation within *Quercus* L. Sect. *Cerris Loudon in Turkey*. *Dendrobiology* **2013**, 109–117. [CrossRef]

57. Jensen, J.S. Variation of Growth in Danish Provenance Trials with Oak (*Quercus robur* L. and *Quercus petraea* Mattuschka Liebl.). *Ann. Des Sci. For.* **1993**, *50*, 203s–207s. [CrossRef]

58. Dormann, C.F.; Elith, J.; Bacher, S.; Buchmann, C.; Carl, G.; Carré, G.; Marquéz, J.R.G.; Gruber, B.; Lafourcade, B.; Leitão, P.J.; et al. Collinearity: A Review of Methods to Deal with It and a Simulation Study Evaluating Their Performance. *Ecography* **2012**, *36*, 27–46. [CrossRef]

59. Chatterjee, S.; Hadi, A.S.; Price, B. Simple Linear Regression. In *Regression Analysis by Example*; John Wiley & Sons, Inc.: Hoboken, NJ, USA, 2006; pp. 21–51.

60. Wickham, H. *Ggplot2: Elegant Graphics for Data Analysis*; Springer: New York, NY, USA, 2016; ISBN 978-3-319-24277-4.

61. Pinheiro, J.; Bates, D.; DebRoy, S.; Sarkar, D.; Team, R.C. Linear and Nonlinear Mixed Effects Models. *R package version* **2007**, *3*, 1–89.

62. Nakagawa, S.; Schielzeth, H. A General and Simple Method for Obtaining R2 from Generalized Linear Mixed-Effects Models. *Methods Ecol. Evol.* **2012**, *4*, 133–142. [CrossRef]

63. Barton, K. MuMIn: Multi-Model Inference. R Package Version 1. 0. 0. 2009. Available online: http://r-forge.r-project.org/projects/mumin/ (accessed on 13 March 2021).

64. Fick, S.E.; Hijmans, R.J. WorldClim 2: New 1-Km Spatial Resolution Climate Surfaces for Global Land Areas. *Int. J. Climatol.* **2017**, *37*, 4302–4315. [CrossRef]

65. Broennimann, O.; Fitzpatrick, M.C.; Pearman, P.B.; Petitpierre, B.; Pellissier, L.; Yoccoz, N.G.; Thuiller, W.; Fortin, M.J.; Randin, C.; Zimmermann, N.E.; et al. Measuring ecological niche overlap from occurrence and spatial environmental data. *Glob. Ecol. Biogeogr.* **2012**, *21*, 481–497. [CrossRef]

66. Schoener, T.W. Nonsynchronous Spatial Overlap of Lizards in Patchy Habitats. *Ecology* **1970**, *51*, 408–418. [CrossRef]

67. Lumley, T. Based on R Code by A. M. Leaps: Regression Subset Selection, R Package Version 3.0. Available online: https://cran.r-project.org/package=leaps (accessed on 10 March 2021).

68. Benjamini, Y.; Hochberg, Y. Controlling the False Discovery Rate: A Practical and Powerful Approach to Multiple Testing. *J. R. Stat. Soc.* **1995**, *57*, 289–300. [CrossRef]

69. Yang, J.-Q.; Hsu, K.-C.; Liu, Z.-Z.; Su, L.-W.; Kuo, P.-H.; Tang, W.-Q.; Zhou, Z.-C.; Liu, D.; Bao, B.-L.; Lin, H.-D. The Population History of *Garra orientalis* (Teleostei: Cyprinidae) Using Mitochondrial DNA and Microsatellite Data with Approximate Bayesian Computation. *BMC Evol. Biol.* **2016**, *16*. [CrossRef]

70. Zhang, X.-W.; Li, Y.; Zhang, Q.; Fang, Y.-M. Ancient East-West Divergence, Recent Admixture, and Multiple Marginal Refugia Shape Genetic Structure of a Widespread Oak Species (*Quercus acutissima*) in China. *Tree Genet. Genomes* **2018**, *14*, 88. [CrossRef]

71. Feng, L.; Zhang, Y.-P.; Chen, X.-D.; Yang, J.; Zhou, T.; Bai, G.-Q.; Yang, J.; Li, Z.-H.; Peng, C.-I.; Zhao, G.-F. Allopatric Divergence, Local Adaptation, and Multiple Quaternary Refugia in a Long-Lived Tree (*Quercus spinosa*) from Subtropical China. *BioRxiv* **2017**. [CrossRef]

72. Navascués, M. MicrosatABC-IM: Approximate Bayesian Computation Inferences under the Isolation with Migration Model from Microsatellite Data. Available online: https://hal.inrae.fr/hal-02789612 (accessed on 20 February 2021). [CrossRef]

73. Hudson, R.R. Generating Samples under a Wright-Fisher Neutral Model of Genetic Variation. *Bioinformatics* **2002**, *18*, 337–338. [CrossRef]

74. Raynal, L.; Marin, J.-M.; Pudlo, P.; Ribatet, M.; Robert, C.P.; Estoup, A. ABC Random Forests for Bayesian Parameter Inference. *Bioinformatics* **2018**, *35*, 1720–1728. [CrossRef]

75. Pudlo, P.; Marin, J.-M.; Estoup, A.; Cornuet, J.-M.; Gautier, M.; Robert, C.P. Reliable ABC Model Choice via Random Forests. *Bioinformatics* **2016**, *32*, 859–866. [CrossRef]

76. Thuillet, A.-C.; Bru, D.; David, J.; Roumet, P.; Santoni, S.; Sourdille, P.; Bataillon, T. Direct Estimation of Mutation Rate for 10 Microsatellite Loci in Durum Wheat, *Triticum turgidum* (L.) Thell. ssp. *Durum* Desf. *Mol. Biol. Evol.* **2002**, *19*, 122–125. [CrossRef]

77. Vigouroux, Y.; Jaqueth, J.S.; Matsuoka, Y.; Smith, O.S.; Beavis, W.D.; Smith, J.S.C.; Doebley, J. Rate and Pattern of Mutation at Microsatellite Loci in Maize. *Mol. Biol. Evol.* **2002**, *19*, 1251–1260. [CrossRef]

78. Marriage, T.N.; Hudman, S.; Mort, M.E.; Orive, M.E.; Shaw, R.G.; Kelly, J.K. Direct Estimation of the Mutation Rate at Dinucleotide Microsatellite Loci in *Arabidopsis thaliana* (Brassicaceae). *Heredity* **2009**, *103*, 310–317. [CrossRef]

79. O'Connell, L.M. Somatic Mutations at Microsatellite Loci in Western Red Cedar (*Thuja plicata*: Cupressaceae). *J. Hered.* **2004**, *95*, 172–176. [CrossRef] [PubMed]

80. Kass, R.E.; Raftery, A.E. Bayes Factors. *J. Am. Stat. Assoc.* **1995**, *90*, 773–795. [CrossRef]

81. Jarosz, A.F.; Wiley, J. What Are the Odds? A Practical Guide to Computing and Reporting Bayes Factors. *J. Probl. Solving* **2014**, *7*, 2. [CrossRef]

82. Ashley, M.V.; Backs, J.R.; Kindsvater, L.; Abraham, S.T. Genetic Variation and Structure in an Endemic Island Oak, *Quercus tomentella*, and Mainland Canyon Oak, *Quercus chrysolepis*. *Int. J. Plant Sci.* **2018**, *179*, 151–161. [CrossRef]

83. Backs, J.R.; Terry, M.; Klein, M.; Ashley, M.V. Genetic Analysis of a Rare Isolated Species: A Tough Little West Texas Oak, *Quercus hinckleyi* C.H. Mull. *J. Torrey Bot. Soc.* **2015**, *142*, 302–313. [CrossRef]

84. Gonzalez-Rodriguez, A.; Oyama, K. Leaf Morphometric Variation in *Quercus affinis* and *Q. laurina* (Fagaceae), Two Hybridizing Mexican Red Oaks. *Bot. J. Linn. Soc.* **2005**, *147*, 427–435. [CrossRef]

85. Kusi, J.; Karsai, I. Plastic Leaf Morphology in Three Species of *Quercus*: The More Exposed Leaves Are Smaller, More Lobated and Denser. *Plant Species Biol.* **2019**, *35*, 24–37. [CrossRef]

86. Trejo, L.; Alvarado-Cárdenas, L.O.; Scheinvar, E.; Eguiarte, L.E. Population Genetic Analysis and Bioclimatic Modeling in *Agave striata* in the Chihuahuan Desert Indicate Higher Genetic Variation and Lower Differentiation in Drier and More Variable Environments. *Am. J. Bot.* **2016**, *103*, 1020–1029. [CrossRef]

87. Sork, V.L.; Davis, F.W.; Westfall, R.; Flint, A.; Ikegami, M.; Wang, H.; Grivet, D. Gene Movement and Genetic Association with Regional Climate Gradients in California Valley Oak (*Quercus lobata* Née) in the Face of Climate Change. *Mol. Ecol.* **2010**, *19*, 3806–3823. [CrossRef]

88. Riordan, E.C.; Gugger, P.F.; Ortego, J.; Smith, C.; Gaddis, K.; Thompson, P.; Sork, V.L. Association of Genetic and Phenotypic Variability with Geography and Climate in Three Southern California Oaks. *Am. J. Bot.* **2016**, *103*, 73–85. [CrossRef]

89. Yang, J.; Vázquez, L.; Feng, L.; Liu, Z.; Zhao, G. Climatic and Soil Factors Shape the Demographical History and Genetic Diversity of a Deciduous Oak (*Quercus liaotungensis*) in Northern China. *Front. Plant Sci.* **2018**, *9*, 9. [CrossRef]

90. Ortego, J.; Gugger, P.F.; Sork, V.L. Climatically Stable Landscapes Predict Patterns of Genetic Structure and Admixture in the Californian Canyon Live Oak. *J. Biogeogr.* **2014**, *42*, 328–338. [CrossRef]

91. Mayes, S.G.; McGinley, M.A.; Werth, C.R. Clonal Population Structure and Genetic Variation in Sand-Shinnery Oak, *Quercus havardii* (Fagaceae). *Am. J. Bot.* **1998**, *85*, 1609–1617. [CrossRef] [PubMed]
92. Hipp, A.L.; Eaton, D.A.R.; Cavender-Bares, J.; Fitzek, E.; Nipper, R.; Manos, P.S. A Framework Phylogeny of the American Oak Clade Based on Sequenced RAD Data. *PLoS ONE* **2014**, *9*, e93975. [CrossRef]
93. Hampe, A.; Petit, R.J. Conserving Biodiversity under Climate Change: The Rear Edge Matters. *Ecol. Lett.* **2005**, *8*, 461–467. [CrossRef]
94. Muller, C.H. Ecological Control of Hybridization in *Quercus*: A Factor in the Mechanism of Evolution. *Evolution* **1952**, *6*, 147. [CrossRef]
95. Hipp, A.L.; Manos, P.S.; Hahn, M.; Avishai, M.; Bodénès, C.; Cavender-Bares, J.; Crowl, A.A.; Deng, M.; Denk, T.; Fitz-Gibbon, S.; et al. Genomic Landscape of the Global Oak Phylogeny. *New Phytol.* **2019**, *226*, 1198–1212. [CrossRef]
96. Lepais, O.; Gerber, S. Reproductive Patterns Shape Introgression Dynamics and Species Succession within the European White Oak Species Complex. *Evolution* **2010**, *65*, 156–170. [CrossRef]
97. Spence, E.S.; Fant, J.; Gailing, O.; Griffith, M.P.; Havens, K.; Hipp, A.L.; Kadav, P.; Kramer, A.; Thompson, P.; Toppila, R.; et al. Comparing genetic diversity in three threatened oaks. *Forests* **2021**, *12*.
98. Chatwin, W.B.; Carpenter, K.K.; Jimenez, F.R.; Elzinga, D.B.; Johnson, L.A.; Maughan, P.J. Microsatellite Primer Development for Post Oak, *Quercus stellata* (Fagaceae). *Appl. Plant Sci.* **2014**, *2*, 1400070. [CrossRef] [PubMed]
99. Dow, B.D.; Ashley, M.V.; Howe, H.F. Characterization of Highly Variable (GA/CT) n Microsatellites in the Bur Oak, *Quercus macrocarpa*. *Theor. Appl. Genet.* **1995**, *91*, 137–141. [CrossRef] [PubMed]
100. Kampfer, S.; Lexer, C.; Glössl, J.; Steinkellner, H. Characterization of (GA) n Microsatellite Loci from *Quercus robur*. *Hereditas* **2004**, *129*, 183–186. [CrossRef]
101. Steinkellner, H.; Lexer, C.; Turetschek, E.; Glössl, J. Conservation of (GA) n Microsatellite Loci between *Quercus* Species. *Mol. Ecol.* **1997**, *6*, 1189–1194. [CrossRef]
102. Isagi, Y.; Suhandono, S. PCR Primers Amplifying Microsatellite Loci of *Quercus myrsinifolia* Blume and Their Conservation between Oak Species. *Mol. Ecol.* **1997**, *6*, 897–899. [CrossRef] [PubMed]

Article

Comparing Genetic Diversity in Three Threatened Oaks

Emma Suzuki Spence [1,*], Jeremie B. Fant [2], Oliver Gailing [3,4], M. Patrick Griffith [5], Kayri Havens [2], Andrew L. Hipp [1], Priyanka Kadav [4], Andrea Kramer [2], Patrick Thompson [6], Raakel Toppila [7], Murphy Westwood [1], Jordan Wood [2], Bethany A. Zumwalde [1,8] and Sean Hoban [1,9,*]

[1] The Morton Arboretum, Center for Tree Science, 4100 Illinois 53, Lisle, IL 60532, USA; ahipp@mortonarb.org (A.L.H.); mwestwood@mortonarb.org (M.W.); bzumwalde@ufl.edu (B.A.Z.)

[2] Negaunee Institute for Plant Conservation Science and Action, Chicago Botanic Garden, 1000 Lake Cook Road, Glencoe, IL 60022, USA; jfant@chicagobotanic.org (J.B.F.); khavens@chicagobotanic.org (K.H.); akramer@chicagobotanic.org (A.K.); woodj.2012@gmail.com (J.W.)

[3] Forest Genetics and Forest Tree Breeding, University of Göttingen, 37077 Göttingen, Germany; ogailin@gwdg.de

[4] College of Forest Resources and Environmental Science, Michigan Technological University, 1400 Townsend Dr, Houghton, MI 49931, USA; pdkadav@mtu.edu

[5] Montgomery Botanical Center, 11901 Old Cutler Rd, Coral Gables, FL 33156, USA; patrick@montgomerybotanical.org

[6] Department of Biological Sciences, Auburn University Davis Arboretum, Auburn, AL 36849, USA; thomppg@auburn.edu

[7] Longwood Graduate Program in Public Horticulture, University of Delaware, 181 South College Avenue, Newark, DE 19717-5267, USA; raakelt@gmail.com

[8] Department of Biology, University of Florida, Gainesville, FL 32611, USA

[9] The Field Museum Chicago, DNA Discovery Center, 1400 S. Lake Shore Dr, Chicago, IL 60605, USA

* Correspondence: spence.suzuki@gmail.com (E.S.S.); shoban@mortonarb.org (S.H.)

Citation: Spence, E.S.; Fant, J.B.; Gailing, O.; Griffith, M.P.; Havens, K.; Hipp, A.L.; Kadav, P.; Kramer, A.; Thompson, P.; Toppila, R.; et al. Comparing Genetic Diversity in Three Threatened Oaks. *Forests* **2021**, *12*, 561. https://doi.org/10.3390/f12050561

Academic Editors: Mary Ashley and Janet R. Backs

Received: 15 March 2021
Accepted: 19 April 2021
Published: 29 April 2021

Publisher's Note: MDPI stays neutral with regard to jurisdictional claims in published maps and institutional affiliations.

Abstract: Genetic diversity is a critical resource for species' survival during times of environmental change. Conserving and sustainably managing genetic diversity requires understanding the distribution and amount of genetic diversity (in situ and ex situ) across multiple species. This paper focuses on three emblematic and IUCN Red List threatened oaks (*Quercus*, Fagaceae), a highly speciose tree genus that contains numerous rare species and poses challenges for ex situ conservation. We compare the genetic diversity of three rare oak species—*Quercus georgiana*, *Q. oglethorpensis*, and *Q. boyntonii*—to common oaks; investigate the correlation of range size, population size, and the abiotic environment with genetic diversity within and among populations in situ; and test how well genetic diversity preserved in botanic gardens correlates with geographic range size. Our main findings are: (1) these three rare species generally have lower genetic diversity than more abundant oaks; (2) in some cases, small population size and geographic range correlate with genetic diversity and differentiation; and (3) genetic diversity currently protected in botanic gardens is inadequately predicted by geographic range size and number of samples preserved, suggesting non-random sampling of populations for conservation collections. Our results highlight that most populations of these three rare oaks have managed to avoid severe genetic erosion, but their small size will likely necessitate genetic management going forward.

Keywords: conservation biology; fragmentation; botanic gardens; EST; inbreeding; heterozygosity; microsatellites; population genetics; ex situ

1. Introduction

Genetic diversity is a critical resource for species to adapt to future challenges including pests and diseases, climate change, and other environmental changes. To conserve and sustainably manage genetic diversity, it is important to understand the distribution and amount of genetic diversity present in situ, and to identify the key factors shaping that genetic diversity [1–3]. While genetic diversity has been assessed in hundreds of rare

97

species globally, there are few comparative, multispecies studies of the influence of range size, environmental, and demographic variables on genetic diversity in rare species. Previous studies have shown that, on average, species with larger range sizes (and thus, more populations, larger populations, and lower genetic drift) have higher genetic diversity [4–6]. Additionally, small local populations whose effective size (N_e) is <500 individuals are generally at risk of genetic drift and loss of adaptive potential; and even smaller populations (<50 individuals) experience rapid genetic erosion and inbreeding problems [7]. Additionally, environmental factors may influence genetic diversity at certain loci through selection by climate or habitat [8,9], or may influence genetic diversity genome-wide through drift as population sizes are reduced or fluctuate in size [10].

It is equally important to understand how well genetic diversity is conserved ex situ. The amount of genetic diversity conserved ex situ may be correlated with some of the same factors impacting genetic diversity in situ (e.g., population size and geographic range size). Species with larger geographic ranges may require more plants ex situ to fully safeguard genetic diversity [11,12], but there are few cases empirically testing this [13]. Determining how many plants are needed to conserve genetic diversity is important to meet global, national, and institutional commitments on conservation. For instance, the Convention on Biological Diversity's Global Strategy for Plant Conservation (GSPC) has created guidelines for crops, crop wild relatives, and other economically important species that calls for "at least 75% of threatened plant species in ex situ collections, preferably in the country of origin", and for "70% of genetic diversity" to be conserved, by 2020 [14]. The GSPC also stipulates that collections be "accessible, backed up, and genetically representative". Collections that meet these requirements provide insurance against extinction, support in situ conservation (e.g., supplementation or reintroduction), and provide material for conservation-relevant research [1,15,16]. Safeguarding exceptional species is especially challenging [17], as they cannot be kept in traditional seed banks but must be kept ex situ in living collections (i.e., of mature individuals rather than seeds). Exceptional species include many common trees such as magnolias, oaks, cycads, and others.

This paper focuses on the tree genus, *Quercus* (oaks), for which there are currently 112 oak species threatened globally under the International Union for Conservation of Nature (IUCN) threat categories vulnerable, endangered, or critically endangered [18] with 17 of those in the United States [19]. Unfortunately, acorns are generally recalcitrant (they cannot be stored in conventional seed banks [20]) and are challenging for tissue culture and cryopreservation [21], making oaks an exceptional species. Conservation through living collections in botanic gardens and arboreta is currently the principal ex situ conservation option for threatened oak species [22,23].

Oaks have significant ecological [24], economic [25], and cultural [26] importance, while facing increasing conservation threats including habitat loss, invasive species, shifting climates, and pests/pathogens, such as sudden oak death [27]. For example, oaks support huge numbers of other species that depend on them for mast or forage [28], and many oak species are used for timber or non-timber forest products [29,30]. Oaks also feature incredible ecological breadth, having diversified into many habitats through adaptations in leaf traits, phenology, habit, water use, and other traits [31,32]. Identifying and quantifying the factors influencing genetic diversity in situ and ex situ is especially important for speciose genera like oaks, that has 450 species estimated worldwide. Characterizing correlates of genetic diversity in a few well-studied species within the oak phylogeny could serve as the foundation for predictive models of the distribution of genetic diversity. Such models should make it possible to manage genetic diversity in a wide range of species, even when data are not available. Discovering the "rules" underlying the distribution of genetic diversity in situ and ex situ can lead to more efficient conservation.

We focus on three threatened oak species that occur in the Southeastern United States, a hotspot for genetic and species richness for plant taxa, including oak species (Figure A1). This region is of particular conservation interest in light of numerous threats including land use change, increasing drought and high temperatures, changing fire regimes, invasive species, urban and suburban sprawl, and extreme weather events [27]. Our focal species for this study—*Quercus boyntonii* Beadle (critically endangered [19]), *Q. oglethorpensis* W.H. Duncan (endangered [19]), and *Q. georgiana* M.A. Curtis (endangered [19])—fit all three categories of rare as outlined by Rabinowitz [33]; small range sizes, high habitat specificity, and low abundance (Table A1). Each species has less than 2000 mature individuals remaining, which are distributed in a few fragmented populations, facing the aforementioned threats. All species are considered habitat specialists (*Q. boyntonii* on sandstone rock outcrops and steep hillsides, *Q. georgiana* on granite rock outcrops, and *Q. oglethorpensis* on poorly drained marshlands or creek bottoms, Table 1), but have differing geographic distributions and habitat requirements (Table 1, Figure 1).

Table 1. Overview and comparison of each of the three species in this study showing geographic range, area of occupancy (AOO), extent of occurrence (EOO), number of adults estimated in the wild, and current population trends. Data compiled from IUCN Red List of Threatened Species.

Species	IUCN Listing; Habitat and Geographic Range	AOO (km^2)	EOO (km^2)	Number of Adult Trees	Population Trend
Quercus boyntonii	Critically Endangered; restricted occurrence in Alabama, restricted to sandstone outcrops and glades, and steep dry hillsides	24	4157	50–200	Stable
Quercus georgiana	Endangered; restricted to isolated granite outcrops and flat-rocks in the Piedmont Plateau of the Southeastern United States.	72–272	16,570–21,600	Unknown	Decreasing
Quercus oglethorpensis	Endangered; disjunct distribution; small clusters of localities in Louisiana, Mississippi, and Alabama, and a more extensive distribution from Northeast Georgia into South Carolina	180–3000	130,000	1000	Decreasing

In these three oak species, we ask whether range size, population size, and environmental variables correlate to genetic diversity within and among populations in situ. We then ask to what degree genetic diversity is preserved in botanic gardens and if this correlates to geographic range size. Specifically, we aim to:

- Assess levels of genetic diversity and differentiation in our focal species and compare them to other rare and common oak species (hypothesis: common oak species have more genetic diversity);
- Determine if genetic diversity and differentiation correlate with the relative range size of the three oak species (hypothesis: larger range size correlates to higher genetic diversity and higher genetic structure)
- Determine if genetic diversity and differentiation are correlated with demographic (e.g., population size) and/or environmental variables (hypothesis: lower genetic diversity in smaller populations; genetic diversity correlates with environmental variables);
- Quantify the amount of genetic diversity of each species that is conserved ex situ and determine whether this is correlated with features of those species (hypothesis: genetic diversity should be predicted by both the number of plants ex situ and the commonness of the species).

With 31% of the United States' oak species now considered of conservation concern, results from this study should assist in assessing genetic diversity of the many other threatened oak species and designing management strategies for them, especially those for which genetic diversity is not available [18].

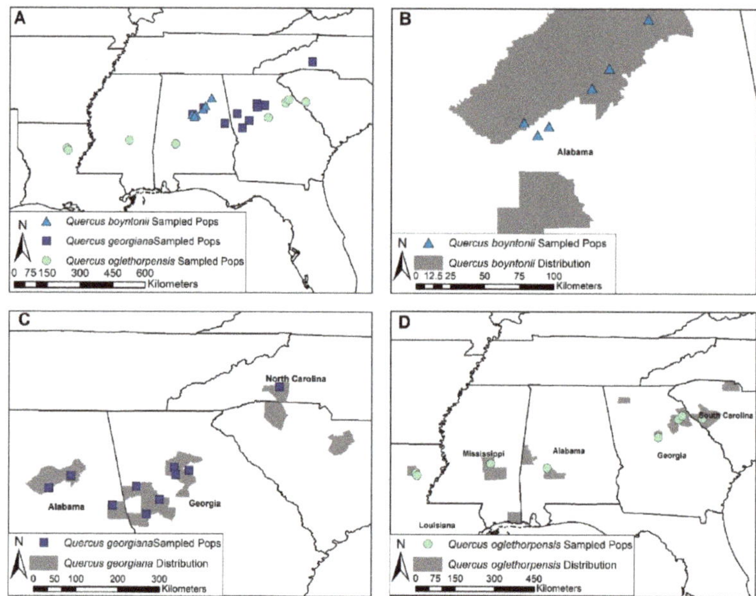

Figure 1. Sampled locations in relation to species geographic range; note that most extant occurrences were sampled. Range is based on county-level occurrences and is derived from USDA PLANTS. Panel (**A**) shows the sampling distribution of all three oak species included in the study; (**B**) depicts the known distribution of *Q. boyntonii* and sampled locations; (**C**) depicts the known distribution of *Q. georgiana* and sampled locations; (**D**) depicts the known distribution of *Q. oglethropensis* and sampled locations.

2. Materials and Methods

2.1. Study Species, Sampling, and Genotyping

These data were collected and used for a previous study answering different questions/in a different application [13]. Previously the data, along with genetic data of eight additional taxa of woody plant, were used to examine genetic diversity ex situ compared to in situ. We briefly described sampling and genotyping methods here for each focal species; complete collection and genotyping methods can be found in Hoban et al. 2020 [13]. For each species, we collected in situ samples from as many known populations as possible, and selected only trees representative of typical leaf morphologies for each focal species to avoid any possible hybrids. However, it is possible some hybrids were sampled because gene flow among oak species can occur (see Discussion). To find ex situ samples, we used Botanic Gardens Conservation International PlantSearch (https://members.bgci.org/data_tools/plantsearch, accessed on 2017), a global database of more than 1000 botanic gardens and their collections.

Quercus boyntonii (Alabama sandstone post oak) is endemic to Alabama (USA.), although historical records say that it formerly grew in Texas [34]. It is a shrub or small tree, sometimes reaching a height of 6 m, but usually smaller. *Q. boyntonii* was sampled ex situ from 16 botanic gardens and arboreta that have *Q. boyntonii* in their collections, totaling 87 individuals. In situ individuals were sampled in natural preserves, private property, and suburban parks. In situ population sizes ranged from fewer than 10 to more than 100 trees. Occurrences of the species are patchy, coinciding with suitable remnant habitat: sandstone outcrops, ridges, and slopes. We sampled 246 in situ samples (227 included in final analysis after clones were removed). In situ samples were collected during May 2017, and ex situ samples were collected between April and September 2017. Due to the patchiness of habitat, occurrence, and wind pollination it is challenging to delimit strict "populations". For these analyses, we used 8 km distance to delineate populations in instances of continuous distributions. We genotyped all individuals using 11 neutral microsatellites from previous studies in oaks. Extraction, testing of the larger panel of markers from which our microsatellites were drawn, and genotyping are discussed in detail in Hoban et al. 2020 [13].

Quercus georgiana (Georgia oak) is native to the Southeastern United States, mainly in Northern Georgia, but with additional populations in Alabama, North Carolina, and South Carolina. It grows on dry granite and sandstone outcrops of slopes of hills at 50–500 meters' altitude [35]. *Quercus georgiana* is a small tree, often shrubby in the wild, growing to 8–15 m tall. *Quercus georgiana* was sampled from nine populations across the known range of the species through the use of herbarium records, collection data from botanic garden records, and USDA PLANTS Database (USDA 2012). All sampled populations were separated by at least 15 km. A total of 226 samples (223 were retained with sufficient genetic data) were sampled in June 2011. At least 24 individual trees were randomly sampled from each site, and sampled plants were at least five meters apart. Seventeen botanical institutions in the United States, France, and Belgium shared samples, totaling 36 individuals. Eight nuclear and 11 expressed sequence tag (EST) microsatellite markers were used for genotyping, following extraction and genotyping methods detailed in Hoban et al. 2020 [13]. We expect EST microsatellites to have lower polymorphism information content as they are associated with transcribed regions of DNA [36]. Nuclear and EST markers were assessed separately for analysis.

Quercus oglethorpensis (Oglethorpe oak) is a long-lived woody plant endemic to the Southeastern United States. Extant and largely fragmented wild populations are documented in South Carolina, Georgia, Alabama, Mississippi, and Louisiana. *Q. oglethorpensis* has a disjunct distribution across its range, with smaller clusters of localities in Northeast Louisiana, Southeast Mississippi, and Southwest Alabama, and a more extensive and well-known distribution from Northeast Georgia across the border into South Carolina. It grows to up to 25 m in height, and has leaves that are flat, narrowly-elliptical and usually without lobes. We prioritized sites with the most up-to-date occurrence data that was gathered in July 2015 during a germplasm collection effort [37]. We included additional sites not visited during the collection effort so that the greatest geographic distribution could be sampled. Sampled populations were separated by at least 9 km. Eight in situ populations were visited for a total of 191 samples (187 were retained with sufficient genetic data). Ex situ samples were collected from 145 trees, representing 16 botanic gardens around the world. All samples were genotyped with 10 nuclear microsatellite markers following extraction and genotyping methods in Hoban et al. 2020 [13].

2.2. Analysis: Basic Statistics

We used the R v 3.6.3 (R Core Team, Vienna, Austria) package adegenet version 2.1.2 to convert genepop files to genind and genpop formats. We used the R package poppr version 2.8.3 [38] to identify potential clones (and to remove clones/duplicate genotypes before any other calculations), expected heterozygosity, number of alleles, and allelic richness; hierfstat version 0.4.22 [39] to calculate pairwise population F_{ST} values; diveRsity version 1.9.9 [40] to calculate observed heterozygosity and inbreeding coefficient (F_{IS}); and Demerelate version 0.9.3 to calculate measures of relatedness [39–44]. We also tested for signatures of recent bottlenecks using the heterozygote excess test in the BOTTLENECK software [45] with both the infinite allele model and the two phase model and the mode shift test. We performed an ANOVA with species as the factor and population as the unit of analysis, to test for differences among species in the main summary statistics. For this and subsequent tests we only used the nuclear SSRs because EST-SSRs have much lower heterozygosity and allelic richness and we only had them for one species. We also tested for isolation by distance with linear regression of genetic distance (F_{ST}) on geographic distance among populations. All analysis scripts are available at https://github.com/smhoban/SE_oaks_genetics (accessed on 15 February 2021).

2.3. Influence of Environment on Genetic Diversity and Differentiation

We obtained 19 standard bioclimatic variables from WorldClim 2.0 at a resolution of 2.5 min [46]. To determine if there is a relationship between local climatic variables and population-level genetics, for each species, we performed ordinary least squares linear regression [47] of each climatic variable on each of four basic population genetic summary statistics that we may expect to respond to local climate: expected heterozygosity, allelic richness, F_{ST}, and relatedness. All analysis scripts are available at https://github.com/smhoban/SE_oaks_genetics (accessed on 15 February 2021).

2.4. Influence of Local Population Size

Following previous work [48–51], we calculated the percentage of genetic diversity conserved as the proportion of extant in situ alleles preserved in ex situ collections. We focused on alleles existing in the wild; we did not count alleles existing only in botanic garden collections. These data were previously presented in Hoban et al. 2020 [13] but were analyzed in a different context: comparing genetic diversity in ex situ collections among different genera and without regard to species range sizes.

3. Results

3.1. Basic Results

Genetic summary statistics for all three species include: N of samples genotyped, genetic diversity, measured as expected heterozygosity (H_e) and allelic richness (A_r), genetic differentiation (pairwise F_{ST}) and relatedness (R), and estimated population size (Table 2). We only present one relatedness estimator [43], but the patterns were similar for all three measures tested. No populations showed significant bottleneck signatures using the heterozygote excess test and the two-phase model, although one population of each species did show a signature of a bottleneck using the heterozygote excess test and the infinite allele model. The smallest population of *Q. oglethorpensis* showed a "mode shift", though no heterozygote excess. For *Q. georgiana* a bottleneck signature was observed for five populations (half of the populations) but only for the EST-SSRs. No bottleneck was detected for *Q. boyntonii* or *Q. georgiana* with neutral microsatellites. *Q. oglethorpensis*, the largest-ranged species we sampled, and was the only species which showed significant isolation by distance.

Comparing these three rare species to a set of other *Quercus* studies, we found that the rare oaks in this study had among the lowest heterozygosity, and that *Q. oglethorpensis* had an exceptionally high inbreeding coefficient (F_{IS}, Table A1).

Table 2. Summary statistics for each population and the average across populations. Reported is the population name (Pop name), the state the population is located (State, specific locality data is not provided given the rarity of the species), the number of samples (*N* samples genotyped) and number of unique multilocus genotypes (unique MLG), the expected heterozygosity (H_{exp}), allelic richness (A_r), mean pairwise F_{ST}, relatedness (Rel), and estimated number of trees based on direct observations in the field (Pop size est.).

Species	Pop Name	State	*N* Samples Genotyped (Unique MLG)	H_{exp}	A_r	Mean Pairwise F_{ST}	Rel	Pop Size est.
	IMLS032	AL	14 (12)	0.581	5.02	0.023	−0.014	20
	IMLS048	AL	17 (15)	0.605	5.45	0.029	0.234	30
	IMLS068	AL	22 (22)	0.605	4.84	0.027	0.024	25
Q. boyntonii	IMLS138	AL	12 (11)	0.593	3.51	0.043	0.013	5
	IMLS280	AL	83 (76)	0.642	5.85	0.015	0.05	165
	IMLS244	AL	60 (60)	0.63	4.78	0.023	0.134	150
	IMLS307	AL	30 (30)	0.651	6.02	0.025	−0.013	70
	AVERAGE	–	34 (32)	0.615	5.07	0.026	0.061	66

Table 2. *Cont.*

Species	Pop Name	State	N Samples Genotyped (Unique MLG)	H_exp	A_r	Mean Pairwise F_ST	Rel	Pop Size est.
	MR24	AL	24	0.738	7.355	0.051	0.031	24
	EDEN32	AL	25	0.624	7.739	0.056	−0.056	50
	Pen32	AL	26	0.736	7.506	0.062	0.156	50
	CB32	GA	24	0.709	6.827	0.049	0.075	100
	DK32	GA	25	0.677	7.503	0.046	0.123	100
Q. georgiana	CR31	GA	24	0.677	6.656	0.056	0.069	50
	SM32	GA	26	0.732	8.114	0.042	0.167	100
	am29	GA	25	0.803	9.706	0.039	0.131	100
	WGHerb	GA	25	0.769	8.993	0.044	0.415	30
	COP-9	NC	26	0.588	5.03	0.074	−0.043	200
	AVERAGE	–	25	0.705	7.54	0.052	0.107	80
	RMS-9	LA	14	0.645	5.9	0.069	0.017	200
	BIE-3-7	MS	33	0.694	6.03	0.052	−0.025	50
Q. oglethorpensis	CAT-9	AL	27	0.639	6.25	0.054	0.075	60
	MOT-9	GA	31	0.653	5.56	0.073	0.172	500
	BUF-9	GA	29	0.62	5.75	0.059	0.008	50
	SUM-9	SC	28	0.648	6.06	0.055	0.028	40
	AVERAGE	–	27	0.65	5.93	0.06	0.046	150

3.2. Genetic Diversity and Range Size for Our Three Rare Oaks

Range size shows some relationship to heterozygosity and allelic richness, in that *Q. boyntonii* (the most geographically restricted species) had the lowest heterozygosity and allelic richness (Figure 2). However, *Q. georgiana* had the highest heterozygosity even though its range size was moderate. Range size strongly related to genetic differentiation as measured by F_{ST}. All ANOVA test comparisons were significant except *Q. georgiana*, *Q. oglethorpensis* for allelic richness and *Q. boyntonii*, *Q. oglethorpensis* for allelic richness and expected heterozygosity.

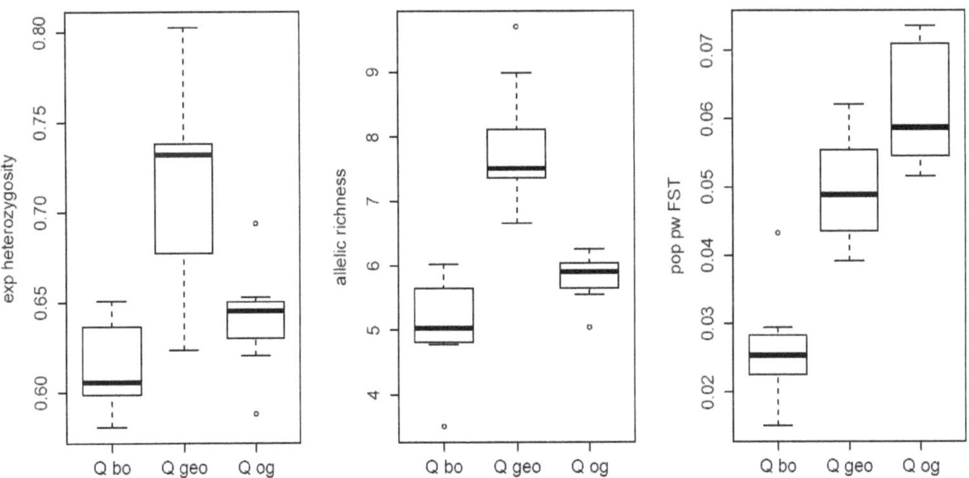

Figure 2. Species genetic diversity boxplots across populations comparing expected heterozygosity (exp heterozygosity), allelic richness, and F_{ST}. (pop pw F_{ST}) Species are arranged by range size, with the smallest range on the left (*Q. boyntonii*, "Q bo"), followed by *Q. georgiana* ("Q geo"), and the largest species range on the right (*Q. oglethorpensis*, "Q og"). Only nuclear SSRs are included in this analysis.

3.3. Genetic Diversity and Range Size within Each Species

Genetic diversity and differentiation statistics are presented in Appendix B for populations above and below N_c of 50 individuals. For two species, *Q. boyntonii* and *Q. georgiana*, the trends were as predicted, with allelic richness and heterozygosity generally higher in larger populations and F_{ST} generally lower for larger populations (Figure A2). Though statistically significant *p*-value differences were observed in only a few comparisons, all other comparisons were not significantly different (Figure A2). Additionally, in these two species, relatedness generally showed no difference. For the third species, *Q. oglethorpensis*, the opposite pattern was observed, with lower genetic diversity, higher differentiation and higher relatedness in larger populations.

3.4. Genetic Diversity and Environment

Genetic diversity and differentiation were not related to climate variables for *Q. georgiana* and *Q. oglethorpensis* when testing all 19 bioclimatic variables at 2.5 min—none were significant after correcting for multiple testing.

3.5. Genetic Diversity in Ex Situ Collections

The percentage of genetic diversity currently preserved in ex situ collections is shown in Table 3. The percentage was not clearly related to species range size or to the number of ex situ samples; although *Q. boyntonii* had the smallest range and moderate number of samples, it had the lowest genetic diversity ex situ. Note that EST diversity was not conserved (Figure A3).

Table 3. Percent allele capture in ex situ collections. The percent of alleles conserved in ex situ collections, for different allele frequency categories, for each species. Allele frequency categories are: all (all alleles); very common alleles (>10%); common alleles (>5%); low (<10% and >1%); and rare (<1%). For rare alleles and all alleles, two results are presented, percentage captured when all alleles including those with fewer than two occurrences are included (complete data), and when alleles with one or two occurrences are excluded (reduced data, shown in parentheses).

Species	N Plants Ex Situ	Geographic Range Size	All	Very Common	Common	Low Frequency	Rare
Q. oglethorpensis	145	large	78 (94)	100	100	97	37 (67)
Q. georgiana (EST-SSR)	36	medium	61 (68)	100	85	51	33 (33)
Q. georgiana (nSSR)	36	medium	69 (76)	100	100	75	35 (41)
Q. boyntonii	77	small	60 (70)	100	100	66	32 (29)

3.6. Other Observations

We only identified clones in *Quercus boyntonii*. For this species we often observed small "rings" or clusters of stems, sometimes 5 or more meters across. We found 12 pairs of clones, which werealways were adjacent individuals, either stems sampled immediately next to each other or within a few meters.

As expected, we found that EST-SSR markers had lower diversity than nuclear SSR markers, with heterozygosity and the number of alleles being 19% and 14% lower on average, respectively.

4. Discussion

Our study tested the influence of range size, environmental and demographic variables on genetic diversity, and differentiation in three rare oak species. Our main findings are as follows. (1) These three rare species generally have lower genetic diversity than more common oaks previously studied, and range size relates strongly to genetic differentiation but less strongly to genetic diversity. (2) In spite of relatively small numbers of populations available, due to the rarity of these species, we found that in some cases small population size and geographic range may correlate with some metrics of genetic diversity and differentiation. (3) We also found that genetic diversity currently conserved varies among species of comparable geographic range size and numbers of samples preserved. Thus, our study supports the idea that "rarity" and collection history are not sufficient to explain genetic diversity in ex situ collections: the amount of genetic diversity preserved is also a function of intrinsic biology, demography, or life histories that vary independently of rarity.

We first present our observations in the context of rare and common species in the genus. Many population genetic studies have been performed in *Quercus* [52,53]. Genetic diversity is often summarized using allelic richness and heterozygosity. Expected heterozygosity was lower in our study (less than 0.65 for most populations, and a mean of 0.641 for *Q. oglethorpensis*, 0.615 for *Q. boyntonii*, and 0.72 for *Q. georgiana*) than was observed in other oaks, which typically had heterozygosities between 0.7 and 0.9 (Table A2). However, some common oaks were observed with lower genetic diversity (e.g., *Q. phillyreoides*, H_e = 0.535) and some rare oaks were observed with higher genetic diversity (e.g., *Q. pacifica* (H_e = 0.851) and *Q. hinckleyi* (H_e = 0.853) as estimated using microsatellites. Some oaks were postulated to be naturally rare (e.g., *Q. boyntonii*), others were more likely to be rare due to human disturbance (e.g., *Q. arkansana*), and others were increasing in rarity for a long time (e.g., *Q. hinckleyi*) [27]. Due to the relatively long-lived nature of most oak species (100+ years), it is possible that recently rare oaks may take a long time to show the subsequent genetic impacts of a drop in population size and narrowing ranges that are associated with their increasing rarity. This form of "extinction debt" has been shown in simulations [12,54], while more naturally rare oaks would not be expected to show such genetic impacts. The relatively low genetic diversity in the species we studied may be due to relatively low population sizes over multiple generations.

Comparing genetic diversity statistics for these three species with different range sizes we see that *Q. boyntonii* has lowest heterozygosity and allelic richness as expected based on small range and highest endangerment status. However, *Q. oglethorpensis* and *Q. georgiana* had relatively equal allelic richness, and *Q. georgiana* had the highest heterozygosity even though its range size was moderate. It is not surprising that overall range size was only a moderate predictor of genetic diversity, as it is the local effective population size that influences retention of genetic diversity within populations (see next section). The paucity of bottleneck signatures may suggest the species have not suffered bottlenecks, or that bottleneck signatures have not had time to develop (as in other species with known, recent population collapses, e.g., *Juglans cinerea*, [55]); bottleneck tests are unreliable for recent, moderate, or gradual bottlenecks [54].

On the other hand, F_{ST} is related to species range size for these three species: the smallest-range species (*Q. boyntonii*) had lowest F_{ST}, and the species with the largest range size and most general habitat preference (*Q. oglethorpensis*) had highest the F_{ST} (Figure 2). This conforms to population genetic theory regarding isolation by distance, whereby populations of a large range species have the most distance among them, and genetic distance is known to increase with geographic distance. Thus, in our study the influence of range size was much more apparent on among population genetic differentiation than on within population genetic diversity. Of course range size is not the only predictor of F_{ST}, factors such as connectivity can also be used to predict F_{ST}. For example, wide ranging oak species with high numbers of populations, and thus high gene flow, can show low F_{ST} (e.g., in *Q. macrocarpa* [56]).

According to conservation genetic theory we would expect that populations near or below a population size of 50 individuals would be subject to strong genetic drift. The exact threshold for a population to rapidly suffer detrimental genetic consequences has been hotly debated [7,57,58], but here we focused on 50 individuals. For our study we would predict lower allelic richness and heterozygosity, and high F_{ST} and relatedness in such populations. We see this predicted pattern in *Q. georgiana* and *Q. boyntonii*, though comparisons were significant or nearly so only for F_{ST} in *Q. georgiana* (all loci *t* test 0.055,

Wilcox 0.063; ESTs *t* test 0.04, Wilcox 0.063) and heterozygosity for *Q. boyntonii* (*t* test 0.003, Wilcox 0.057). The relatively low number of significant values emphasizes the small number of replicate populations (inherent in rare species) and the fact that for very recently reduced populations, genetic diversity impacts may not yet have accumulated [55].

Interestingly, for *Q. oglethorpensis* all statistics are in the opposite direction of what might be predicted based on a large population size (higher F_{ST}, higher relatedness, lower heterozygosity, and lower allelic richness). It is not clear why *Q. oglethorpensis* shows this pattern. This could be a result of fragmentation coupled with the fact that *Q. oglethorpensis* grows predominantly as a subcanopy tree [59]. Although not well-studied in wind pollinated trees, subcanopy habit could possibly limit pollen dispersal [60]. However, this pattern would be consistent with recent expansions or founding populations, which would result in moderate population size but reduced genetic diversity and increased F_{ST}.

There was no relationship between environmental variables and genetic statistics. It is possible that for these species, neutral genetic diversity is more influenced by current population sizes, which may be impacted by processes other than environment, such as land development, loss of habitat, etc. It is also possible that neutral genetic diversity and demography are influenced by environment but at fine spatial scales and/or along unmeasured environmental axes. Useful future work will be to create ecological niche models for each species to test for the impact of habitat suitability/probability of occurrence in relation to genetic diversity [61–63]. All three species are habitat specialists with typically very restricted populations.

Although the very common and just common alleles are preserved well in ex situ collections, low frequency and rare alleles are not conserved well, and overall, only a moderate amount of genetic diversity is preserved ex situ, between 60 and 78% for these species assuming all alleles are considered (68–94% if the rarest alleles are dropped). Previous modeling work suggests that the species with the largest range and highest F_{ST} would require the most samples [11,12]. *Quercus oglethorpensis* is preserved extremely well, at 78%, even though it has the largest range; it does have the most individuals ex situ. Less of the genetic variation of *Q. boyntonii* is conserved than of *Q. georgiana*, even though *Q. boyntonii* has the smaller range and about twice the number of individuals ex situ. Other studies of the genetic diversity conserved ex situ have primarily been species specific and we are only aware of a few attempts to determine if genetic diversity ex situ correlates to range size. In the plant genus: *Zelkova*, Christe et al. [64] found that a small-range endemic was less well conserved than a larger-range species. Several reasons can explain their similar findings: for the rare species, collectors may have revisited a single accessible site for seed collection, even though it occurred across high topographic and ecological diversity, while for the common species collectors in multiple countries had visited numerous populations. In other words, accessibility and availability of sampling are important to consider.

Although more than 3000 botanical institutions maintain more than 100,000 globally threatened species ex situ [65], the conservation value of these collections is unclear. Most taxa are held in a small number of collections, usually with a small number of inadequately documented accessions [66,67]. While some collections maintain relatively high levels of genetic diversity [68,69], research on the genetic representativeness of species in living collections is sparse. Our results emphasize that the genetic diversity conserved in collections is not only a function of the number of samples conserved, nor simply a function of the species inherent characteristics such as range size. Rather, the amount of genetic diversity conserved is likely a function of the interaction number of samples, range size, and collection strategy (such as which populations are visited, the spatial sampling within populations, the number of maternal plants collected from, etc.) [51,70]. While *Q. oglethorpensis* is conserved quite well, *Q. boyntonii* and *Q. georgiana* may need more individuals sampled to better represent in situ diversity.

Caveats

We used microsatellites because they are an affordable method to achieve an understanding of genetic diversity and structure. We recognize that increased resolution could be obtained with next generation sequencing techniques [71,72]. It is known that microsatellites that are developed in one species and applied in a different species can show reduced genetic diversity due to PCR amplification failure caused by mutations in primer binding sites. The markers applied to our species were all developed from other species of oaks, but were developed in *Quercus* subgenus *Quercus* sections to which they were applied (red oak markers from Section *Lobatae* for the one red oak species, and white oak markers from Section *Quercus* for the two white oak species). However, this is also the case for nearly all microsatellite studies of oaks: the majority of microsatellites were developed in European white oaks and then applied in diverse species (Table A2). It is also known that microsatellites are susceptible to ascertainment bias, such that the investigator will select markers that are polymorphic in a small sample of test individuals, such that less polymorphic markers are not included in the study. We did not have an a priori expectation that the patterns we saw between species were due to this reason, as this should apply to all oak species using these markers.

Other caveats involve the populations we studied. There are likely some populations of these species that we are unaware of, and we sometimes were not able to collect all of the individuals within a population. Moreover, genetic diversity in some populations may reflect gene flow from other species. It is known that gene flow among related oak species does occur, often at low levels and that hybridization may be even higher in species that have low population numbers due to the phenomenon of pollen swamping, where heterospecific pollen may far outnumber conspecific pollen [73,74]. For instance, in the extremely rare *Q. hinckleyi*, hybrids have been identified with genetic markers [52]. We did attempt to only sample individuals consistent with the phenotype of the target species. Of course, any of these caveats would obscure the patterns that we were testing for, and it is possible that if such caveats could be taken into account (for example identifying and removing all hybrids), the patterns we found here might be stronger.

5. Conclusions and Conservation Implications

We found that genetic diversity and differentiation were influenced by both population size and range size, but that patterns did not perfectly accord to predictions. This emphasizes stochastic processes and the influence of multiple factors on genetic diversity we see today (time, human influence, and population recovery). We also found that genetic diversity conserved ex situ was not well predicted by species geographic range size or number of samples, in contrast to theoretical predictions, and that two species need more samples ex situ. The overall low genetic diversity in these three rare oaks relative to more common oaks suggest that genetic diversity may also be low in other threatened oak species, a supposition to be tested by analyzing several more threatened oaks. We note that *Q. oglethorpensis*, in spite of its wide geographic range, had lower allelic richness and heterozygosity than might be expected—nearly as low as the critically endangered and small range *Q. boyntonii*—and thus might already be suffering genetic erosion in its isolated populations.

Author Contributions: Conceptualization: E.S.S., J.B.F., O.G., M.P.G., K.H., A.K., A.L.H., M.W., J.W. and S.H.; methodology: E.S.S., J.B.F., O.G., B.A.Z. and S.H.; software: E.S.S. and S.H.; validation: E.S.S., J.B.F., O.G., M.P.G., K.H., A.L.H., J.W., B.A.Z. and S.H.; formal analysis: E.S.S., P.K., R.T., J.W. and S.H.; investigation: E.S.S., J.B.F., O.G., M.P.G., K.H., A.L.H., P.K., A.K., P.T., R.T., M.W., J.W., B.A.Z. and S.H.; resources, M.P.G., K.H., M.W. and S.H.; data curation, S.H.; writing—original draft preparation: E.S.S. and S.H.; writing—review and editing, E.S.S., J.B.F., O.G., M.P.G., K.H., A.L.H., B.A.Z. and S.H.; visualization: E.S.S. and S.H.; supervision, J.B.F., K.H., O.G., M.W. and S.H.; project administration: J.B.F., M.P.G., K.H. and S.H.; funding acquisition: J.B.F., M.P.G., K.H., A.L.H., A.K., P.T., M.W., J.W. and S.H. All authors have read and agreed to the published version of the manuscript.

Funding: This research was funded by the institute of Museum and library Services, grant number MG-30-16-0085-16, and also in part by the Center for Tree Science at the Morton Arboretum.

Institutional Review Board Statement: Not applicable.

Data Availability Statement: The data and code for analyses used in this study are openly available on the Github repository: https://github.com/smhoban/SE_oaks_genetics (accessed on 15 February 2021).

Acknowledgments: This was truly a group effort, with so many wonderful collaborators to thank. The authors would like to thank all permit-granting agencies, all institutions who voluntarily sent samples, and all the individuals and institutions who aided in collection of wild material. We would like to thank Cindy Johnson, Kevin Feldheim, and Isabel Distefano for volunteering their time and expertise. Additionally, we thank the Field Museum for providing us access to equipment for microsatellite analysis.

Conflicts of Interest: The authors declare no conflict of interest. The funders had no role in the design of the study; in the collection, analyses, or interpretation of data; in the writing of the manuscript, or in the decision to publish the results.

Appendix A

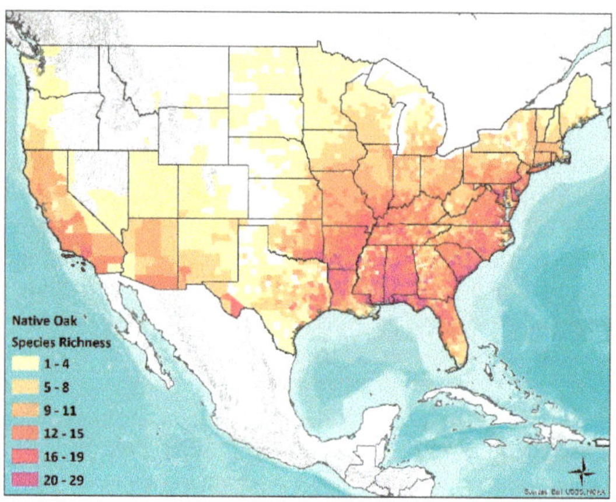

Figure A1. Native U.S. oak species richness by county. County level distribution data from USDA PLANTS and Biota of North America Program (BONAP) were combined to estimate species richness.

Table A1. The matrix of rarity as presented by Rabinowitz mapped with three rare oaks (*Q. boyntonii, Q. georgiana,* and *Q. oglethorpensis*) and three oaks which are common by two measures (geographic range and habitat) but rare in abundance (*Q. hemisphaerica, Q. incana,* and *Q. laevis*). The three rare species group together in the same rarity ranking, while the three common oaks group together in a different rarity ranking. Bold font indicates species is in the subgenus *Erythrobalanus* (red oak), regular font indicates species is in the subgenus *Leucobalanus* (white oak).

Geographic Range	Large		Small	
Habitat Specificity	Wide	Narrow	Wide	Narrow
Local Population Size				
Large, Dominate somewhere				
Small, non-dominant	*Q. hemisphaerica* *Q. incana* *Q. laevis*			*Q. boyntonii* *Q. georgiana* Q. oglethorpensis

Table A2. Literature review and comparison of previously published population genetic studies of oak species and species accessed in this study.

Species	Section	IUCN Status	AOO	EOO	# of Loci	# of Pops	Sample Size (N)	F_{ST}	G_{ST}	Mean Number of Alleles	AR	H_o	H_e	F_{IS}	Citation
Quercus phillyraeoides	*Cerris*	*NEN*	NR	NR	11	24	536	0.097	0.09	10.45	4.47	0.51	0.54	0.079	[75]
Quercus georgiana	*Lobatae*	*EN*	272	21,600	8	9	224	0.049	NR	8.03	7.82	0.72	0.69	0.019	-
Quercus rubra	*Lobatae*	*LC*	NR	4,150,115	10	23	980	0.044	NR	NR	5.5– 12.14	NR	NR	NR	[76]
Quercus berberidifolia	*Quercus*	*LC*	NR	250,000	8	2	60	NR	NR	18.6	5.72	0.85	0.88	0.039	[52]
Quercus boyntonii	*Quercus*	*CR*	24	4,157	9	7	238	0.026	NR	7.24	5.07	0.62	0.55	0.061	-
Quercus dumosa	*Quercus*	*EN*	620	12,500	8	2	24	NR	NR	11.1	4.9	0.77	0.8	0.029	[77]
Quercus hinckleyi	*Quercus*	*CR*	30	380	8	4	123	NR	0.03	13.06	5.15–14.73	0.81	0.85	0.036	[52]
Quercus lobata Née	*Quercus*	*NT*	NR	280,000	8	12	270	0.064	NR	8.12	NR	0.8	0.7	-0.196	[78]
Quercus macrocarpa	*Quercus*	*LC*	NR	4056,799	5	14	480	0.027	NR	11.184	NR	0.92	0.86	NR	[79,80]
Quercus oglethorpensis	*Quercus*	*EN*	3000	130,000	9	7	188	0.062	NR	6.81	5.8	0.64	0.48	0.234	-
Quercus pacifica	*Quercus*	*EN*	NR	3800	8	3	133	0.233	NR	15.8	5.54	0.81	0.85	0.05	[77]
Quercus chrysolepis	*Protobalanus*	*LC*	NR	920,000	8	7	100	NR	NR	16.4	4.63	0.71	0.83	0.146	[53]
Quercus tomentella	*Protobalanus*	*EN*	250	43,500	8	6	345	NR	NR	16.3	3.63	0.56	0.76	0.262	[53]
AVERAGE of previously studied species			300	1,079,677	8	10	305	0.093	0.06	13.45	4.82	0.75	0.79	0.056	
AVERAGE of all species			699	822,738	8	9	285	0.075	0.06	11.92	5.29	0.72	0.73	0.069	

Definitions: (F_{ST}): fixation index; (G_{ST}): Nei's estimation F_{ST} generalized for multiple alleles; (AR): Allelic richness; (H_O): observed heterozygosity; (H_E): expected heterozygosity; (F_{IS}): inbreeding coefficient; (NR): Not Reported. The IUCN status abbreviations are as follows: (NE): Not Evaluated; (LC): Least Concern; (NT): Near Threatened; (EN): Endangered; (CR): Critically Endangered.

Appendix B

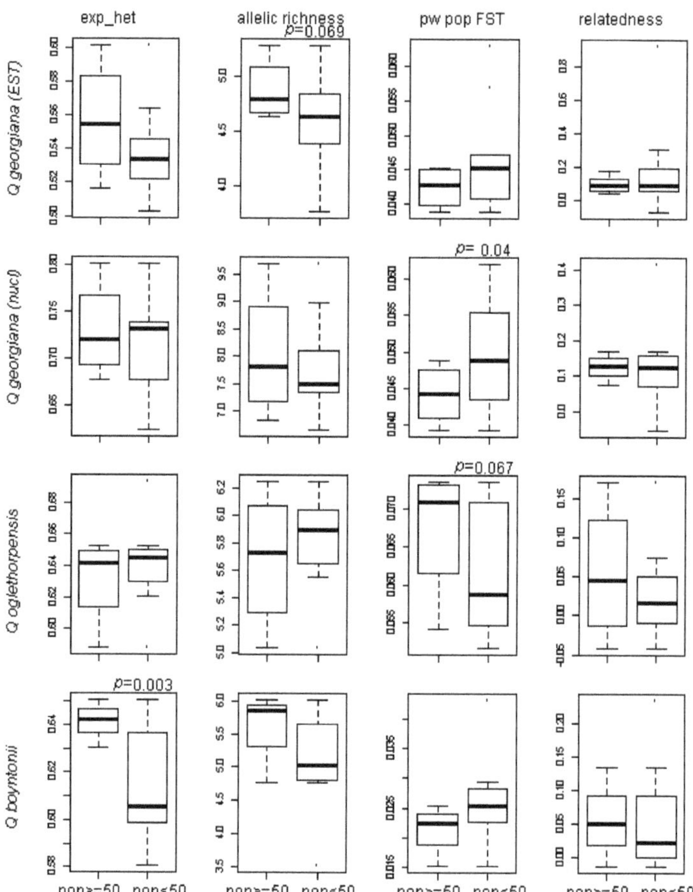

Figure A2. Box and whisker plots for each species by genetic summary. Each column of graphs represents different genetic summary statistics: (exp_het): expected heterozygosity; allelic richness; (pw pop FST): Pairwire population F_{ST}; and relatedness. Within each graph, the boxplot on the right represents populations with greater than or equal to 50 individuals (pop >= 50) and less than 50 individuals (pop < 50).

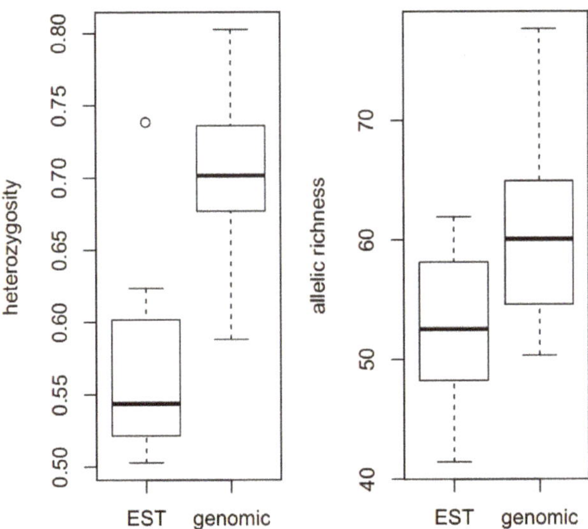

Figure A3. Box and whisker plots comparing heterozygosity and allelic richness between EST and genetic markers for the same species/samples.

References

1. Guerrant, E.O.; Havens, K.; Vitt, P. Sampling for Effective Ex Situ Plant Conservation. *Int. J. Plant Sci.* **2014**, *175*, 11–20. [CrossRef]
2. Namoff, S.; Husby, C.E.; Francisco-Ortega, J.; Noblick, L.R.; Lewis, C.E.; Griffith, M.P. How well does a botanical garden collection of a rare palm capture the genetic variation in a wild population? *Biol. Conserv.* **2010**, *143*, 1110–1117. [CrossRef]
3. Whitlock, R.; Hipperson, H.; Thompson, D.B.A.; Butlin, R.K.; Burke, T. Consequences of in-situ strategies for the conservation of plant genetic diversity. *Biol. Conserv.* **2016**, *203*, 134–142. [CrossRef]
4. Frankham, R. Relationship of Genetic Variation to population size in wildlife. *Conserv. Biol.* **1996**, *10*, 1500–1508. [CrossRef]
5. Pelletier, T.A.; Carstens, B.C. Geographical range size and latitude predict population genetic structure in a global survey. *Biol. Lett.* **2018**, *14*, 20170566. [CrossRef]
6. Hamrick, J.L.; Godt, M.J.W. Effects of life history traits on genetic diversity in plant species. *Philos. Trans. R. Soc. Lond. B Biol. Sci.* **1996**, *351*, 1291–1298. [CrossRef]
7. Frankham, R.; Bradshaw, C.J.A.; Brook, B.W. Genetics in conservation management: Revised recommendations for the 50/500 rules, Red List criteria and population viability analyses. *Biol. Conserv.* **2014**, *170*, 56–63. [CrossRef]
8. Kremer, A.; Potts, B.M.; Delzon, S. Genetic divergence in forest trees: Understanding the consequences of climate change. *Funct. Ecol.* **2014**, *28*, 22–36. [CrossRef]
9. Sork, V.L.; Davis, F.W.; Westfall, R.; Flint, A.; Ikegami, M.; Wang, H.; Grivet, D. Gene movement and genetic association with regional climate gradients in California valley oak (*Quercus lobata* Née) in the face of climate change. *Mol. Ecol.* **2010**, *19*, 3806–3823. [CrossRef]
10. Eckert, C.G.; Samis, K.E.; Lougheed, S.C. Genetic variation across species' geographical ranges: The central–marginal hypothesis and beyond. *Mol. Ecol.* **2008**, *17*, 1170–1188. [CrossRef]
11. Hoban, S. New guidance for ex situ gene conservation: Sampling realistic population systems and accounting for collection attrition. *Biol. Conserv.* **2019**, *235*, 199–208. [CrossRef]
12. Hoban, S.; Schlarbaum, S. Optimal sampling of seeds from plant populations for ex-situ conservation of genetic biodiversity, considering realistic population structure. *Biol. Conserv.* **2014**, *177*, 90–99. [CrossRef]
13. Hoban, S.; Callicrate, T.; Clark, J.; Deans, S.; Dosmann, M.; Fant, J.; Gailing, O.; Havens, K.; Hipp, A.L.; Kadav, P.; et al. Taxonomic similarity does not predict necessary sample size for ex situ conservation: A comparison among five genera. *Proc. R. Soc. B* **2020**, *287*, 20200102. [CrossRef] [PubMed]
14. Jackson, P.W.; Kennedy, K. The Global Strategy for Plant Conservation: A challenge and opportunity for the international community. *Trends Plant Sci.* **2009**, *14*, 578–580. [CrossRef]
15. Cavender, N.; Westwood, M.; Bechtoldt, C.; Donnelly, G.; Oldfield, S.; Gardner, M.; Rae, D.; McNamara, W. Strengthening the conservation value of *ex situ* tree collections. *Oryx* **2015**, *49*, 416–424. [CrossRef]
16. Westwood, M.; Cavender, N.; Meyer, A.; Smith, P. Botanic garden solutions to the plant extinction crisis. *Plants People Planet* **2021**, *3*, 22–32. [CrossRef]

17. Pence, V.C. In Vitro methods and the challenge of exceptional species for target 8 of the global strategy for plant conservation1. *Ann. Mo. Bot. Gard.* **2013**, *99*, 214–220. [CrossRef]
18. Carrero, C.; Jerome, D.; Beckman, E.; Byrne, A.; Coombes, A.J.; Deng, M.; González-Rodríguez, A.; Hoang, V.S.; Khoo, E.; Nguyen, N.; et al. *The Red List of Oaks 2020*; The Morton Arboretum: Lisle, IL, USA, 2020.
19. Jerome, D.; Beckman, E.; Kenny, L.; Wenzell, K.; Kua, C.-S.; Westwood, M. *US Oaks*; The Morton Arboretum: Lisle, IL, USA, 2017.
20. Connor, K.F. Update on oak seed quality research: Hardwood recalcitrant seeds. National Proceedings, forestry and conser-vation nursery associations—2003. Ogden (UT): USDA Forest Service. *Rocky Mt. Res. Stn. Proc. RMRS-P-33* **2004**, 111–116.
21. Kramer, A.T.; Pence, V. The challenges of *ex situ* conservation for threatened oaks. *Intl. Oaks* **2012**, *23*, 91–108.
22. Fant, J.B.; Havens, K.; Kramer, A.T.; Walsh, S.K.; Callicrate, T.; Lacy, R.C.; Maunder, M.; Meyer, A.H.; Smith, P.P. What to do when we cant bank on seeds: What botanic gardens can learn from the zoo community about conserving plants in living collections. *Am. J. Bot.* **2016**, *103*, 1541–1543. [CrossRef]
23. Wood, J.; Ballou, J.D.; Callicrate, T.; Fant, J.B.; Griffith, M.P.; Kramer, A.T.; Lacy, R.C.; Meyer, A.; Sullivan, S.; Traylor-Holzer, K.; et al. Applying the zoo model to conservation of threatened exceptional plant species. *Conserv. Biol.* **2020**, *34*, 1416–1425. [CrossRef] [PubMed]
24. Tallamy, D.W.; Shropshire, K.J. Ranking Lepidopteran use of native versus introduced plants. *Conserv. Biol.* **2009**, *23*, 941–947. [CrossRef]
25. Varela, M.C.; Eriksson, G. Multipurpose gene conservation in *Quercus suber*—A Portuguese example. *Silvae Genet.* **1995**, *44*, 28–36.
26. Leroy, T.; Plomion, C.; Kremer, A. Oak symbolism in the light of genomics. *New Phytol.* **2019**, *226*, 1012–1017. [CrossRef] [PubMed]
27. Beckman, E. *Conservation Gap Analysis of Native US Oaks*; Morton Arboretum: Lisle, IL, USA, 2019.
28. McShea, W.J.; Healy, W.M.; Devers, P.; Fearer, T.; Koch, F.H.; Stauffer, D.; Waldon, J. Forestry matters: Decline of oaks will impact wildlife in hardwood forests. *J. Wildl. Manag.* **2007**, *71*, 1717–1728. [CrossRef]
29. Cavender-Bares, J. Diversity, distribution and ecosystem services of the North American oaks. *Int. Oaks* **2016**, *27*, 37–48.
30. Logan, W.B. *Oak: Frame of Civilization*; W.W. Norton & Company, Inc.: New York, NY, USA, 2005.
31. Cavender-Bares, J. Diversification, adaptation, and community assembly of the American oaks (*Quercus*), a model clade for integrating ecology and evolution. *New Phytol.* **2019**, *221*, 669–692. [CrossRef]
32. Leroy, T.; Louvet, J.; Lalanne, C.; Le Provost, G.; Labadie, K.; Aury, J.; Delzon, S.; Plomion, C.; Kremer, A. Adaptive introgression as a driver of local adaptation to climate in European white oaks. *New Phytol.* **2020**, *226*, 1171–1182. [CrossRef]
33. Synge, H. *The Biological Aspects of Rare Plant Conservation Edited by Hugh Synge, Seven forms of rarity*; John Wiley & Sons Ltd.: New York, NY, USA, 1981; pp. 205–217.
34. Beckman, E.; Hoban, S.; Spence, E.; Meyer, A.; Westwood, M. *Species Profile: Quercus boyntonii Beadle*; Conservation Gap Analysis of Native U.S: Thousand Oaks, CA, USA, 2019; pp. 74–79.
35. Toppila, R. Ex Situ Conservation of Oak (*Quercus L.*) in Botanic Gardens: A North American Perspective. Ph.D. Thesis, University of Delewarw, Newark, DE, USA, 2012.
36. Chabane, K.; Ablett, G.A.; Cordeiro, G.M.; Valkoun, J.; Henry, R.J. EST versus genomic derived microsatellite markers for genotyping wild and cultivated barley. *Genet. Resour. Crop. Evol.* **2005**, *52*, 903–909. [CrossRef]
37. Lobdell, M.S.; Thompson, P.G. Ex-situ Conservation of *Quercus Oglethorpensis* in Living Collections of Arboreta and Botanical Gardens. In *Gene Conservation of Tree Species—Banking on the Future. Proceedings of a Workshop*; Sniezko, R.A., Man, G., Hipkins, V., Woeste, K., Gwaze, D., Kliejunas, J.T., McTeague, B.A., Eds.; *Tech. Cords. Gen. Tech. Rep.* PNW-GTR-963; Department of Agriculture, Forest Service Pacific Northwest Research Station: Portland, OR, USA, 2017; Volume 963, pp. 144–153.
38. Kamvar, Z.N.; Tabima, J.F.; Grünwald, N.J. Poppr: An R package for genetic analysis of populations with clonal, partially clonal, and/or sexual reproduction. *PeerJ* **2014**, *2*, e281. [CrossRef]
39. Goudet, J. Hierfstat, a package for r to compute and test hierarchical F-statistics. *Mol. Ecol. Notes* **2005**, *5*, 184–186. [CrossRef]
40. Keenan, K.; McGinnity, P.; Cross, T.F.; Crozier, W.W.; Prodöhl, P.A. diveRsity: An R package for the estimation and explo-ration of population genetics parameters and their associated errors. *Methods Ecol. Evol.* **2013**, *4*, 782–788. [CrossRef]
41. Kraemer, P.; Gerlach, G. Demerelate: Calculating interindividual relatedness for kinship analysis based on codominant diploid genetic markers using R. *Mol. Ecol. Resour.* **2017**, *17*, 1371–1377. [CrossRef]
42. Blouin, M.S.; Parsons, M.; Lacaille, V.; Lotz, S. Use of microsatellite loci to classify individuals by relatedness. *Mol. Ecol.* **1996**, *5*, 393–401. [CrossRef] [PubMed]
43. Ritland, K. Estimators for pairwise relatedness and individual inbreeding coefficients. *Genet. Res.* **1996**, *67*, 175–185. [CrossRef]
44. Wang, J. An estimator for pairwise relatedness using molecular markers. *Genetics* **2002**, *160*, 1203–1215. [CrossRef] [PubMed]
45. Piry, S.; Luikart, G.; Cornuet, J.-M. Bottleneck: A program for detecting recent effective population size reductions from allele data frequencies. *J. Hered.* **1999**, *90*, 502–503. [CrossRef]
46. Fick, S.E.; Hijmans, R.J. WorldClim 2: New 1-km spatial resolution climate surfaces for global land areas. *Int. J. Climatol.* **2017**, *37*, 4302–4315. [CrossRef]
47. R Core Team. *R: A Language and Environment for Statistical Computing*; R Core Team: Vienna, Austria, 2020.
48. Petit, R.J.; Hampe, A. Some Evolutionary Consequences of Being a Tree. *Annu. Rev. Ecol. Evol. Syst.* **2006**, *37*, 187–214. [CrossRef]
49. Wojacki, J.; Eusemann, P.; Ahnert, D.; Pakull, B.; Liesebach, H. Genetic diversity in seeds produced in artificial Douglas-fir (*Pseudotsuga menziesii*) stands of different size. *For. Ecol. Manag.* **2019**, *438*, 18–24. [CrossRef]

50. Marshall, D.R.; Brown, A.H.D. Optimum Sampling Strategies in Genetic Conservation. In Crop Genetic Resources for Today and Tomorrow. Cambridge University Press: Cambridge, UK, 1975; pp. 53–80.

51. Hoban, S.; Strand, A. *Ex situ* seed collections will benefit from considering spatial sampling design and species' reproductive biology. *Biol. Conserv.* **2015**, *187*, 182–191. [CrossRef]

52. Backs, J.R.; Terry, M.; Ashley, M.V. Using genetic analysis to evaluate hybridization as a conservation concern for the threatened species *Quercus hinckleyi* C.H. Muller (Fagaceae). *Int. J. Plant Sci.* **2016**, *177*, 122–131. [CrossRef]

53. Ashley, M.V.; Backs, J.R.; Kindsvater, L.; Abraham, S.T. Genetic variation and structure in an endemic island oak, *Quercus tomentella*, and mainland Canyon oak, *Quercus chrysolepis*. *Int. J. Plant Sci.* **2018**, *179*, 151–161. [CrossRef]

54. Hoban, S.M.; Gaggiotti, O.E.; Bertorelle, G. The number of markers and samples needed for detecting bottlenecks under realistic scenarios, with and without recovery: A simulation-based study. *Mol. Ecol.* **2013**, *22*, 3444–3450. [CrossRef] [PubMed]

55. Hoban, S.M.; Borkowski, D.S.; Brosi, S.L.; McCLEARY, T.S.; Thompson, L.M.; McLACHLAN, J.S.; Pereira, M.A.; Schlarbaum, S.E.; Romero-Severson, J. Range-wide distribution of genetic diversity in the North American tree *Juglans cinerea*: A prod-uct of range shifts, not ecological marginality or recent population decline. *Mol. Ecol.* **2010**, *19*, 4876–4891. [CrossRef]

56. Schnabel, A.; Hamrick, J.L. Comparative Analysis of Population Genetic Structure in *Quercus macrocarpa* and *Q. gambelii* (Fagaceae). *Syst. Bot.* **1990**, *15*, 240. [CrossRef]

57. Franklin, I.R.; Frankham, R. How large must populations be to retain evolutionary potential? *Anim. Conserv.* **1998**, *1*, 69–70. [CrossRef]

58. Hastings, K.; Frissell, C.A.; Allendorf, F.W. Naturally isolated coastal cutthroat trout populations provide empirical support for the 50/500 rule. In Proceedings of the 2005 Coastal Cutthroat Trout Symposium, Port Townsend, Washington, DC, USA, 29 September–1 October 2005; p. 121.

59. Stein, J.D. *Field Guide to Native Oak Species of Eastern North America*; U.S. Forest Service, Forest Health Technology Enterprise Team: Morgantown, WV, USA, 2001.

60. Koenig, W.D.; Ashley, M.V. Is pollen limited? The answer is blowin' in the wind. *Trends Ecol. Evol.* **2003**, *18*, 157–159. [CrossRef]

61. Ortego, J.; Riordan, E.C.; Gugger, P.F.; Sork, V.L. Influence of environmental heterogeneity on genetic diversity and structure in an endemic southern Californian oak. *Mol. Ecol.* **2012**, *21*, 3210–3223. [CrossRef]

62. Lira-Noriega, A.; Manthey, J.D. Relationship of genetic diversity and niche centrality: A survey and analysis. *Evolution* **2014**, *68*, 1082–1093. [CrossRef]

63. Diniz-Filho, J.A.F.; Rodrigues, H.; Telles, M.P.D.C.; De Oliveira, G.; Terribile, L.C.; Soares, T.N.; Nabout, J.C. Correlation between genetic diversity and environmental suitability: Taking uncertainty from ecological niche models into account. *Mol. Ecol. Resour.* **2015**, *15*, 1059–1066. [CrossRef]

64. Christe, C.; Kozlowski, G.; Frey, D.; Fazan, L.; Bétrisey, S.; Pirintsos, S.; Gratzfeld, J.; Naciri, Y. Do living *ex situ* collections capture the genetic variation of wild populations? A molecular analysis of two relict tree species, *Zelkova abelica* and *Zelkova carpinifolia*. *Biodivers. Conserv.* **2014**, *23*, 2945–2959. [CrossRef]

65. Sharrock, S.; Oldfield, S.; Wilson, O. *Plant Conservation Report 2014: A Review of Progress towards the Global Strategy for Plant Conservation 2011-2020. CBD Technical Series*; Convention on Biological Diversity: Montreal, QC, Canada, 2014.

66. Hurka, H. Conservation Genetics and the Role of Botanical Gardens. In *Conservation Genetics*; Loeschcke, V., Jain, S.K., Tomiuk, J., Eds.; Birkhäuser Basel: Basel, Switzerland, 1994; pp. 371–380. ISBN 9783034885102.

67. Maunder, M.; Higgens, S.; Culham, A. The effectiveness of botanic garden collections in supporting plant conservation: A European case study. *Biodivers. Conserv.* **2001**, *10*, 383–401. [CrossRef]

68. Griffith, M.P.; Calonje, M.; Meerow, A.W.; Tut, F.; Kramer, A.T.; Hird, A.; Magellan, T.M.; Husby, C.E. Can a botanic garden Cycad collection capture the genetic diversity in a wild population? *Int. J. Plant Sci.* **2015**, *176*, 1–10. [CrossRef]

69. Sharrock, S.; Jones, M. Saving Europe's threatened flora: Progress towards GSPC Target 8 in Europe. *Biodivers. Conserv.* **2010**, *20*, 325–333. [CrossRef]

70. Hoban, S.; Volk, G.; Routson, K.J.; Walters, C.; Richards, C. Sampling Wild Species to Conserve Genetic Diversity. In *North American Crop Wild Relatives*; Springer: Berlin/Heidelberg, Germany, 2018; pp. 209–228.

71. Hodel, R.G.J.; Segovia-Salcedo, M.C.; Landis, J.B.; Crowl, A.A.; Sun, M.; Liu, X.; Gitzendanner, M.A.; Douglas, N.A.; Germain-Aubrey, C.C.; Chen, S.; et al. The Report of my death was an exaggeration: A review for researchers using microsatellites in the 21st century. *Appl. Plant Sci.* **2016**, *4*, 1600025. [CrossRef]

72. Angeloni, F.; Wagemaker, N.; Vergeer, P.; Ouborg, J. Genomic toolboxes for conservation biologists. *Evol. Appl.* **2011**, *5*, 130–143. [CrossRef]

73. Petit, R.J.; Bodénès, C.; Ducousso, A.; Roussel, G.; Kremer, A. Hybridization as a mechanism of invasion in oaks. *New Phytol.* **2004**, *161*, 151–164. [CrossRef]

74. Petit, R.J.; Latouche-Hallé, C.; Pemonge, M.-H.; Kremer, A. Chloroplast DNA variation of oaks in France and the influence of forest fragmentation on genetic diversity. *For. Ecol. Manag.* **2002**, *156*, 115–129. [CrossRef]

75. Harada, K.; Dwiyanti, F.; Liu, H.-Z.; Takeichi, Y.; Nakatani, N.; Kamiya, K. Genetic variation and structure of Ubame oak, *Quercus phillyraeoides*, in Japan revealed by chloroplast DNA and nuclear microsatellite markers. *Genes Genet. Syst.* **2018**, *93*, 37–50. [CrossRef]

76. Borkowski, D.S.; Hoban, S.M.; Chatwin, W.; Romero-Severson, J. Rangewide population differentiation and population substruc-ture in *Quercus rubra* L. *Tree Genet. Genomes* **2017**, *13*, 67. [CrossRef]

77. Backs, J.R.; Ashley, M.V. Evolutionary history and gene flow of an endemic island oak: *Quercus pacifica*. *Am. J. Bot.* **2016**, *103*, 2115–2125. [CrossRef] [PubMed]
78. Ashley, M.V.; Abraham, S.T.; Backs, J.R.; Koenig, W.D. Landscape genetics and population structure in Valley Oak (*Quercus lobata* Née). *Am. J. Bot.* **2015**, *102*, 2124–2131. [CrossRef] [PubMed]
79. Craft, K.J.; Ashley, M.V. Landscape genetic structure of bur oak (*Quercus macrocarpa*) savannas in Illinois. *For. Ecol. Manag.* **2007**, *239*, 13–20. [CrossRef]
80. Dow, B.D.; Ashley, M.V. Microsatellite analysis of seed dispersal and parentage of saplings in bur oak, *Quercus macrocarpa*. *Mol. Ecol.* **1996**, *5*, 615–627. [CrossRef]

Review

Use of Genomic Resources to Assess Adaptive Divergence and Introgression in Oaks

Desanka Lazic [1], Andrew L. Hipp [2], John E. Carlson [3] and Oliver Gailing [1,4,*]

1 Department of Forest Genetics and Forest Tree Breeding, Georg-August University of Göttingen, 37007 Göttingen, Germany; desanka.lazic@uni-goettingen.de
2 Center for Tree Science, The Morton Arboretum, Lisle, IL 60532, USA; ahipp@mortonarb.org
3 The Schatz Center for Tree Molecular Genetics, Pennsylvania State University, University Park, State College, PA 16802, USA; jec16@psu.edu
4 Center for Integrated Breeding Research (CiBreed), Georg-August University of Göttingen, 37073 Göttingen, Germany
* Correspondence: ogailin@gwdg.de

Citation: Lazic, D.; Hipp, A.L.; Carlson, J.E.; Gailing, O. Use of Genomic Resources to Assess Adaptive Divergence and Introgression in Oaks. *Forests* **2021**, *12*, 690. https://doi.org/10.3390/f12060690

Academic Editors: Mary Ashley and Janet R. Backs

Received: 30 April 2021
Accepted: 24 May 2021
Published: 27 May 2021

Abstract: Adaptive divergence is widely accepted as a contributor to speciation and the maintenance of species integrity. However, the mechanisms leading to reproductive isolation, the genes involved in adaptive divergence, and the traits that shape the adaptation of wild species to changes in climate are still largely unknown. In studying the role of ecological interactions and environment-driven selection, trees have emerged as potential model organisms because of their longevity and large genetic diversity, especially in natural habitats. Due to recurrent gene flow among species with different ecological preferences, oaks arose as early as the 1970s as a model for understanding how speciation can occur in the face of interspecific gene flow, and what we mean by "species" when geographically and genomically heterogeneous introgression seems to undermine species' genetic coherence. In this review, we provide an overview of recent research into the genomic underpinnings of adaptive divergence and maintenance of species integrity in oaks in the face of gene flow. We review genomic and analytical tools instrumental to better understanding mechanisms leading to reproductive isolation and environment-driven adaptive introgression in oaks. We review evidence that oak species are genomically coherent entities, focusing on sympatric populations with ongoing gene flow, and discuss evidence for and hypotheses regarding genetic mechanisms linking adaptive divergence and reproductive isolation. As the evolution of drought- and freezing-tolerance have been key to the parallel diversification of oaks, we investigate the question of whether the same or a similar set of genes are involved in adaptive divergence for drought and stress tolerance across different taxa and sections. Finally, we propose potential future research directions on the role of hybridization and adaptive introgression in adaptation to climate change.

Keywords: *Quercus*; ecological speciation; genetic mosaic of speciation; introgression; reproductive isolation; species concepts

1. Introduction

Darwin's [1] account of the role of natural selection in speciation is arguably the most unifying principle in evolutionary biology. Yet remarkably, the mechanisms by which it may lead to reproductive isolation in the face of ongoing gene flow and the affected genes are still largely unknown [2–5]. Ecological speciation—the evolution of reproductive isolation between populations by adapting to different environments [2]—has in the past two decades become widely recognized as an important source of species diversity [6–10]. This perspective was accepted by many evolutionary biologists, building off the biological species concept even before evidence mounted in support of ecological speciation; but increasing evidence for environmental selection as separable from other divergence mechanisms [10] has rendered the concept of ecological speciation a unifier for

115

speciation drivers as diverse as climate, resource competition, and predation [10]. It has become increasingly clear that the early focus on allopatric speciation, in which post-zygotic incompatibilities evolved in separated populations with no or limited gene flow [11,12], accounts for only a portion of the diversity of life [3,13]. To know the tree of life fully requires studying barriers to gene flow in populations that are not completely diverged and not yet reproductively isolated, before genetic changes contributing to reproductive isolation become confounded with other differences that accumulate after speciation [3].

One difficulty in studying genetic signatures of sympatric speciation has been technical: studying the maintenance of species integrity in the face of gene flow relies strongly on identifying enough molecular markers to tease apart the effects of genes that are driving divergence from those that are homogenized by gene flow [2,14–16]. Leaning on this concept of speciation in the face of gene flow, Via & West [17] coined the term "genetic mosaic of speciation", based on evidence that divergent selection affects the genes of adaptive key traits while the rest of the genome can remain similar between species [18,19]. The related metaphor of "genomic islands of divergence" describes genomic heterogeneity in differentiation resulting from divergent selection [8,20]. Genomic regions under direct selection—the "islands"—or loci linked to them are expected to show signatures of relatively high interspecific differentiation in comparison to the genome as a whole. These "islands" form and then grow in size due to "divergence hitchhiking" [3,17], a reduction in gene flow in regions adjacent to selected genes that allows alleles involved in reproductive isolation to accumulate in the face of interspecific gene flow [3,8,21]. Due to divergence hitchhiking and a genomically localized reduction in interspecific gene flow, linkage disequilibrium is expected to be higher in those regions that are under divergent selection when compared to control regions [3,14,15,17].

While the role of environmental selection in speciation is widely recognized, the genetic mechanisms linking adaptive divergence and reproductive isolation are still unclear [2,22–24]. The majority of adaptive divergence studies have been performed in model organisms [25,26]. Only a few "speciation" genes, especially causing hybrid inviability and sterility, have been identified, and mainly in model species [8,13,27–29]. As a consequence, little is known about the locations and distribution of regions/loci involved in adaptive divergence and reproductive isolation [30], especially in plants [24,29,31].

The high genetic diversity in tree species, especially in their natural habitats, makes them particularly well suited to research into adaptive divergence in response to changing environments [32]. There are other advantages to choosing trees as models for the study of genome-wide adaptive variation, including their limited history of domestication; their large, open-pollinated native populations [33]; their predominantly random mating systems; and their large effective population sizes [34]. Advances in next-generation sequencing technologies and bioinformatics in the last 15 years have greatly impacted our understanding of forest tree diversity and biology [35]. Tree genomics has benefited from the sequencing of several whole genomes [36–41], which have enabled studies of evolutionary history [32] and the potential roles of tens of thousands of genes on diversification, adaptation, and tree biology generally [35,38].

Oaks have long been recognized as important models for understanding ecological controls on speciation and gene flow, as edaphic and climatic factors shape patterns of distribution and introgression [42–46]. This may make oaks particularly well suited to understanding the genomic mosaic of introgression. A recent population genomic study using whole-genome sequencing in oaks has confirmed that divergence between hybridizing species is limited to a subset of the genome, maintaining species integrity in the divergent portions while the rest of the genome is permeable to gene flow: high levels of differentiation ($F_{ST} > 0.8$) were found at narrow regions distributed across the genome with an average width of 10 kb, indicating that selection in these regions counteracts the homogenizing effect of gene flow [47]. In this context, a question of interest to our understanding of the longevity and persistence of species is whether species evolved in allopatry or sympatry, and whether and in which period there was potential gene flow between them, including whether introgression has been ongoing through the lifespan of the species or initiated after secondary contact [48]. Because divergent selection in strictly allopatric species proceeds unaffected by homogenizing effects of gene flow, the genetic mosaic is less pronounced in species that have diverged in allopatry [8]. In this respect, too, oaks present themselves as an ideal study subject, as the high diversity of oaks growing in sympatry across most of North America and much of western Europe and east Asia serve as replicated natural experiments in oak diversification. At an estimated 435 species, *Quercus* is one of the most diverse and most widespread woody plant genera in the northern hemisphere [44,49–52], and whole genomes of *Quercus robur* [53] and *Q. lobata* [54], as well as draft genome of *Q. suber* [41] position oaks as a model genus for understanding the evolution of reproductive isolation and divergence in the face of gene flow.

2. Oaks as a Model to Study Adaptive Divergence between Species

Oaks are important both economically and ecologically, not just for human use, but also as a shelter and food source for wild animals [55]. As far back as the colonization of early *Homo sapiens* of the Middle East and Europe [56], oaks are entangled in human history, symbolism, tradition, economics, and livelihoods [57–61]. Acorn findings in caves suggest that oaks fed humans in their early settlements and have long served as both a "famine food" and a chosen food, even when other options are available [59,62]. Oaks have also been valuable in cultural and religious life, from association with important gods (e.g., [58]) to becoming a national symbol for such modern countries as Germany, the UK, Poland, Portugal, and, in 2004, the USA [31]. In a review by Leroy et al. [31], the symbolism of oak trees and their significance for humans was connected with attributes of the oak genome. For example, longevity, cohesiveness, and robustness of oaks are three pillars on which the symbolism of oaks stands. As a genetic consequence of longevity, accumulation of heritable mutations is expected. However, Leroy et al. [31] summarized recent findings [38,63] that independently reported only small numbers of somatic mutations (17 and 46 single-nucleotide polymorphisms) at the whole-genome level. Their robustness was supported by Plomion et al. [38], revealing patterns of immune system diversification with an expansion of resistance (R) genes accounting for 9% of the gene catalogue [31]. Thus, the importance of oaks to humans is deeply rooted in their diversification history and biology.

Oaks have been described as a thorny problem [31] and a "worst case scenario" for the biological species concept [13], and they have been included among "botanical horror" taxa [64]. Oaks are problematic for classical biological or morphological species concepts because of weak interspecific boundaries and high intraspecific variation relative to interspecific differentiation [65–70]. Due to these observations, it has been suggested that speciation in oaks may be driven by ecological factors, as reflected in their occurrence in different micro-environments, rather than by strong post-zygotic reproductive isolation [71,72]. Oaks, however, are not unique in the difficulties they pose for delimiting species in light of hybridization. They do exhibit introgression, but many other taxa do as well. Perhaps more important is the fact that oaks are particularly well studied and extremely rich in species, with many occurring in sympatry and thus subject to multispecies introgression [42,73]. Oaks are, for these reasons, among the most famous syngameons (groups of species that maintain their distinctions despite sympatry and regular introgressive hybridization [73–76]).

3. The Reality of Oak Species and Oak Introgression

Species boundaries in oaks are often genetically and morphologically ambiguous [55,66,77,78]. Still, different species typically form genetically disjunct clusters [66,78–83] and maintain ecological distinctions, with ecologically divergent oaks commonly growing together in the same forest but in different micro-environments [84–87]. Oaks hybridize readily within sections, but hybrids between different sections are exceptionally rare in botanical gardens and have not been recorded in nature [55,88,89], though ancient introgression has been detected between sections *Quercus* and *Ponticae* [90] and sections *Quercus* and *Protobal-*

anus [91,92]. Good examples of species pairs that occur in sympatry but differ in local adaptations (for example to drought) include *Q. rubra* and *Q. ellipsoidalis* in North American red oaks (section *Lobatae*), *Q. robur* and *Q. petraea* in European white oaks (section *Quercus*) [93–95], and *Q. virginiana* and *Q. geminata* in the North American live oaks (section *Virentes*) [96]. These and many other species pairs are known to hybridize where their ranges overlap [70,77,97,98], with introgression shifting over time in many cases as climate and habitats change [45,46,99]. Assessing levels of hybridization is necessary in studying the effects of interspecific gene flow and divergent selection of species [100].

Until the 1980s, hybrids were characterized purely based on morphological characteristics such as leaf morphology [67,101–106], and later based on genetic markers that helped distinguishing species of the same taxonomic section [81,100,107,108]. Some studies have combined morphometric and molecular approaches to identify hybrids [68,109–111]. While understanding morphological and functional variation in combination with molecular variation is key to understanding how introgression shapes species evolution and ecology, molecular genetic markers are more useful for identifying hybrids and quantifying introgression, because molecular markers generally show higher discriminatory power and accuracy in species assignment than morphological characteristics and directly sample the genomic background that is most often the target of studies.

Hybridization rate and introgression can be estimated based on identification of hybrids and introgressive forms across species ranges or in sympatric stands using genetic assignment analyses in adult trees, seeds or seedlings; or using paternity analysis, which is limited to sympatric stands [89]. In either case, analyses are based on species-discriminating markers (fixed or nearly fixed within species; e.g., [75,112] or markers polymorphic in all species (e.g., [113])). Several studies of contemporary interspecific gene flow using paternity analysis have been conducted in multi-species stands of European white oaks [98,114,115]. A comparatively high level of interspecific gene flow and introgression was found in the contact zone of species in each example. Using a set of six polymorphic microsatellite loci and paternity analysis in 320 acorns from four oak species, Curtu et al. [98] found an overall hybridization rate of 35.9%, with first generation hybrids present at a rate of 8.4%. Similarly, by means of paternity analysis, a comparable rate of F_1 hybrids (9.0%) was found in 623 acorns by Lepais and Gerber [114] using 10 microsatellite markers in four oak species. Comparable overall proportions of hybrids between *Q. petraea* and *Q. pubescens* (26%) were revealed by paternity analysis of acorns by Salvini et al. [115]. Microsatellite markers were used also in contemporary gene flow studies in North American oaks [70,77]. Paternity analysis of *Q. coccinea*, *Q. rubra*, *Q. velutina* and *Q. falcata* demonstrated more than 20% of seedlings in two mixed-species stands to be potential hybrids [70]. Interspecific gene flow was also confirmed by Khodwekar and Gailing [77] based on parentage analysis of 466 seeds collected from 15 genetically assigned seed parents in sympatric stands of *Q. rubra* and *Q. ellipsoidalis*, which revealed that 3.9% of the seedlings (0–25% for individual seed parents) were derived from interspecific matings.

Genetic assignment analyses are more commonly used to estimate introgression rates. Analyzing five nuclear microsatellite markers with both multivariate discriminant and Bayesian clustering methods implemented in STRUCTURE [116], Valbuena-Carabana et al. [87] quantified genetic differentiation and introgression in 176 adult trees of *Q. petraea* and *Q. pyrenaica*. They found reliable minimum estimates of introgression between species to be relatively low, identifying only 8.5% of adult trees as putative first and later generation hybrids. Similar results were found in a Romanian oak stand comprising the four interfertile species *Q. petraea*, *Q. robur*, *Q. pubescens*, and *Q. frainetto* [80]. Using genetic assignment analysis in adult trees, introgression rates between pairs of species varied, with the highest value between *Q. pubescens* and *Q. frainetto* (16.2%) and the lowest value between *Q. robur* and *Q. frainetto* (1.7%). Comparatively low percentages of adult trees (2.1%–4.6%) and acorns (1.5%) with hybrid ancestry were found in mixed stands of *Q. lobata* and *Q. douglasii* [108,117]. While hybrids are often comparatively frequent in direct contact zones of interfertile species, they rarely occur in pure stands (e.g., [81]). Despite this preponderance of gene-conduits in the contact zones of species, we still observe long-term maintenance of species. A range-wide study of four interfertile oak species—*Q. ellipsoidalis*, *Q. coccinea*, *Q. rubra*, and *Q. velutina*—demonstrated that genetically intermediate individuals or putative F_1 hybrids were present in contact zones between species, where they were inferred to act as a bridge for recurrent gene flow [83]. The same four species were studied by Zhang et al. [118] using maternally inherited chloroplast DNA markers for the first time in red oaks. At three chloroplast microsatellite markers, a total of 23 haplotypes were observed. In all cases, neighboring interspecific population pairs shared haplotypes, indicating contemporary interspecific gene flow. Using a set of 58 unlinked SNPs from coding regions, Reutimann et al. [65] assessed the extent of admixture in three presumably mixed populations of the white oak species *Q. robur*, *Q. petraea*, and *Q. pubescens* across Switzerland. The Bayesian STRUCTURE method and a machine learning approach (support vector machine) recovered admixture levels of 35–36%. Levels of admixture varied across the species pairs, but the highest degree was detected in mixed stands of *Q. petraea* and *Q. pubescens*, indicating high levels of gene flow between these two species. Since no fixed alleles were recovered in any of the taxa, levels of interspecific heterozygosity could not be quantified. This may be a general issue in oaks, in which pure stands (lacking admixture) and fixed alleles are nearly impossible to find (though cf. [112], in which apparently fixed SNPs were identified using RAD-seq and then used to genotype range-wide samples in [75]).

The weight of evidence thus demonstrates two important facts. First, oak species cohere genetically. When DNA data are directed at oaks, for the most part they are found to cluster in genetic space (going back to the foundational study of [119]). The published exceptions we are aware of (e.g., [70,120,121]) involve relatively small numbers of microsatellites, and studies using larger numbers of markers on the same taxa (e.g., [66,75]) have generally found the species to separate in molecular genotypic space. Second, numerous studies (cited in paragraphs above) demonstrate that introgression is ongoing at a fairly high rate, and consequently that conduits for gene flow between species may be present on the landscape at rates of 9–20% or higher, depending on the species pair. Assuming that these rates are representative of the long-term histories of gene flow between oak species, these two findings in combination strongly suggest that oaks cohere genetically in spite of ongoing gene flow. Thus, the data support the existence of a syngameon of interconnected oak species, evolving both separately and together, as hypothesized in the mid-1970s [71,72].

4. Evidence for Selection against Hybrids

While the number of hybrids and introgressive forms seems to be low in adult trees [80,81,87,89], gene flow analyses suggest frequent interspecific gene flow [70,78,98,115], which may suggest selection against hybrids from seedling to adult stage. Thus, Curtu et al. [98] found a much higher proportion of hybrids in the seed generation as determined by paternity analysis (35.9%) when compared to the adult trees (20.1%) in a multi-species oak stand, suggesting selection against many or perhaps most hybrid genotypes may be a mechanism for the maintenance of species identity. On the other hand, Abraham et al. [117] found low numbers of hybrids in both adults (2.1%) and acorns (1.5%) in *Q. lobata* and *Q. douglasii* stands. It is important to note that these observational studies cannot distinguish between selection and stand-level historical changes in species composition and structure, climate, or management that might result in inter-generational differences in introgression rate. While they are strongly suggestive, direct evidence of selection against hybrids would require the assessment of hybrids and parental species over time, ideally from seed to juvenile and adult stage. To our knowledge, such long-term experiments have not been performed yet. In their absence, the sum of observational studies in independent mixed-oak stands showing consistently higher introgression rates in the acorn/seedling generation relative to the adults nonetheless gives credence to the hypothesis that selection against most but not all hybrid genotypes contributes to the observed genomic heterogeneity of introgression observed in adult trees.

Experimental evidence for selection against hybrids was provided by the comparison of non-germinated but viable seeds and germinated seedlings from a hybrid zone between *Q. rubra* and *Q. ellipsoidalis* in a greenhouse common garden experiment [122]. Using genetic assignment analyses at microsatellite markers, Gailing and Zhang [122] characterized and compared the number of hybrids and "pure" species between non-germinated acorns and seedlings of two interfertile sympatric species, *Q. rubra* and *Q. ellipsoidalis*. Hybrids and introgressive forms in non-germinated acorns showed a frequency of 43.6% while they had a much lower frequency of 9.3% in the seedlings, indicating early selection against hybrids possibly as result of intrinsic incompatibilities between species [122]. However, direct evidence for environment dependent selection against hybrids is still missing.

5. Distribution of Hybrids and Species Indicates Environmental Selection

Curtu et al. [80,98] found in a multi-species stand in central Romania that the white oak species *Q. robur*, *Q. petraea*, *Q. pubescens*, and *Q. frainetto* were distributed according to their soil preferences and environmental requirements, and that interspecific hybrids occurred in environmentally intermediate contact zones between species. This result is in accordance with several other findings in oaks [46,70,77,86] and a long history of expectations about the effect of "hybridization of the habitat" on distribution of plant hybrids [123,124]. For example, Khodwekar & Gailing [77] found that the distribution of species and hybrids was significantly associated with soil properties (i.e., soil water holding capacity, nutrient availability). Additionally, closely related oaks have been found to co-occur less frequently at the plot scale than distantly related oaks due to convergence on traits under strong environmental filtering [84,125]. The rapid evolution of fine-scale habitat differentiation seems, in fact, to be a hallmark of oak diversification [126] that contributes to the separation of potentially introgressing species both among and within forest stands.

These results suggest that pre-zygotic [97,100,114] and post-zygotic [98,122] isolation mechanisms play complementary roles in the maintenance of species identity in oaks. Findings of Gailing and Zhang [122] together with species distribution according to environmental requirements and observed lower hybrid frequencies in ancient hybrid zones [127,128] serve as indirect evidence for selection as an important post-zygotic isolation mechanism in oaks. Experimentally disentangling pre-zygotic and post-zygotic isolation mechanisms might include (1) observations of the number of produced seeds and seedling survival derived from intra- and interspecific crosses, accompanied by (2) an assessment of hybrid survival in transplant experiments. Significantly lower hybrid survival as compared to "pure" species in each parental environment but higher survival in intermediate environments would indicate environmental selection (Figure 1). Alternatively, lower hybrid survival across all three environments would be indicative of selection against hybrids as result of intrinsic incompatibilities between species. While field studies are most appropriate for directly estimating selective effects, the axes of environmental variation affecting hybrid performance in intermediate and parental environments might be distinguished under controlled greenhouse conditions, for example with contrasting watering regimes. Whatever their form, experimental studies are an important next step toward a more nuanced understanding of the role of selection in oak species cohesion.

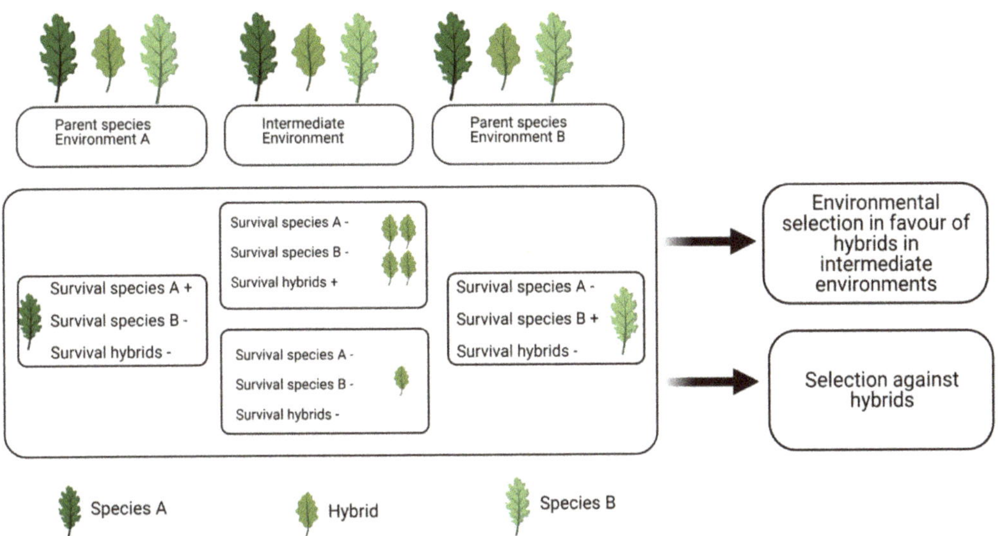

Figure 1. Outline of suggested transplant experiment. Hybrids and parental species are planted in each parental environment and in an intermediate environment. Environmental selection suggests the highest survival rate of parent species A and B in their respective environments. Two different scenarios may be expected for hybrids. Scenario 1: Hybrids show low survival in each parental environment, but higher survival in the intermediate environment, suggesting environmental selection in favor of hybrids in ecological transition zones. Scenario 2: Hybrids show low survival in each of the three environments, suggesting selection against hybrids. Created with BioRender.com, accessed on April 1st 2021.

6. Adaptive Introgression

The introgression of potentially adaptive alleles can have a large impact on adaptation to different environmental conditions. Evidence for adaptive introgression of outlier alleles was provided by the comparison of hybrid numbers and outlier allele frequencies in sympatric and parapatric populations of *Q. rubra* and *Q. ellipsoidalis* [77,129]. In neighbouring parapatric populations of the two species, genetic assignment analysis indicated symmetric interspecific gene flow, but introgression of *Q. ellipsoidalis*-specific alleles at a species-discriminating outlier locus (*CONSTANS*-like 1) into *Q. rubra* occurred at a lower rate than the introgression of *Q. rubra*-specific alleles into *Q. ellipsoidalis* [129]. Sympatric populations in dry outwash plains exhibited a similar discrepancy between symmetrical genome-scale introgression and asymmetrical outlier-locus introgression. However, outlier introgression from the drought adapted species *Q. ellipsoidalis* into *Q. rubra* was more prevalent in these dry outwash plains [77], concordant with the predictions of adaptive introgression [130]. Different directions of outlier alleles introgression in sympatric and parapatric populations suggests shifting selection regimes on these outliers in different spatial and environmental contexts. As these two species have different environmental preferences and outlier allele introgression seems to be related to environmental contrasts, the maintenance of genetic differences between these two species is likely dependent at least in part on environmental selection [77].

Clinal variation consistent with the geographic variation of introgression from *Q. robur* into *Q. petraea* displayed by some genes associated with phenological divergence in *Q. petraea* indicated adaptive introgression in outlier regions [44]. *Quercus robur*-like alleles, for example SNPs located in two important genes controlling stomatal responses, *RopGEF1* and *PBL10* (=*APK1b*) [131,132], were found with higher frequency in *Q. petraea* populations growing at higher altitudes in cooler and/or wetter environments, indicating that introgressed alleles from *Q. robur* serve as a source of phenotypic and potentially adaptive variation in *Q. petraea* [44]. These findings mirror and move beyond the findings of an earlier study that used range-wide genome scans of AFLPs in Californian red oaks [42] to show that the movement of introgressed loci beyond the natural range of their source species was predicted by climate, but without identifying the genes involved. With ongoing climate change, we may expect to find additional opportunities to study adaptive introgression. It is anticipated, for example, that Mediterranean oaks (*Q. pyrenaica* and *Q. pubescens*) will migrate north into the distribution area of temperate oak species (*Q. petraea* and *Q. robur*), providing a chance to study evolutionary processes in action at the present time [24]. Beginning these studies now may give us the opportunity to understand the roles of hybridization and adaptive introgression in climate change adaptation [47].

7. Oak Reference Genome Sequences

A full appreciation of all of the genetic components of variation underlying the capacity for adaptive divergence, speciation, reproductive isolation, and individual quantitative adaptive traits ultimately requires high-quality genome sequence assembly references [133]. Reference chromosome-scale genomes support investigations into variations in the coding sequence, chromosome structure at the macro (whole-genome) and micro (indel) levels, regulatory elements, non-coding RNAs, and positional effects (such as repetitive DNA environs). The approach initially available for assembling draft genome references was the de novo assembly of contigs and scaffolds, which for oak species included pedunculate oak, *Q. robur* (17,910 scaffolds) [53], and valley oak, *Q. lobata* (18,512 scaffolds) [54], both in the *Quercus* section of the genus and *Q. suber* (cork oak, (23,347 scaffolds) [41]) in the section *Cerris*. The *Q. robur* de novo assembly scaffolds were then anchored (ordered) by reference to a high-density genetic linkage map [134], providing chromosome-scale sequences for *Q. robur* covering 96% (716.6 Mb) of the haploid genome size [38]. The contiguous chromosome-scale genome for *Q. robur* contains 90% of the consensus set of 25,808 annotated protein-coding genes predicted in the de novo genome. Certain features of oak genomes, discovered from the chromosome-scale genome assembly of *Q. robur*, may impact studies on hybridization, introgression, and adaptation, including a high level (35.6%) of proximal tandem duplications of genes and inheritance of somatic mutations arising during the long lifespan of oak trees [38]. The extent to which tandem duplication of genes is consistent across the genus, or specific to lineages within oaks, will require chromosome-scale genome assemblies for representatives of the other sections of the oak genus. Recent technology advancements for enabling chromosome-scale genome assemblies, such as Hi-C and nanopore, are now widely in use. A contiguous chromosome-scale genome assembly for northern red oak (*Q. rubra*), in the *Quercus* section *Lobatae*, has recently been completed and will soon be released on the Joint Genome Institute's Phytozome 13 (https://phytozome-next.jgi.doe.gov/, accessed on 28 April 2021). The northern red oak genome, and anticipated additional chromosome-scale oak genomes for oaks, will enable detailed comparative studies at the whole-genome level on pan-genome organization, phylogeny, maintenance of species integrity, hybridization, and adaptations among oaks genus-wide.

8. Genomic Distribution and Architecture of Differentiation

As already pointed out, selection may direct interspecific gene flow between sympatric oak species. However, outliers exist because effective gene flow is in at least some loci reduced, potentially accelerating the accumulation of alleles involved in reproductive isolation [21]. Genome scans in both European white oaks (section *Quercus* in part) and North American red oaks (section *Lobatae*), each of which is composed of ecologically divergent but interfertile oak species, have revealed strong divergent selection on only a few so-called outlier loci [14,44,79,135,136]. Outlier loci exhibit high allele-frequency divergence and are presumed in many cases to be affected by selection in the genomic region in which they occur, due either to direct selection or selection on closely linked loci. They are referred to as "outliers" because they exhibit differentiation that deviates significantly from neutral expectations [136]. Methods that are widely used to detect these loci are F_{ST}-based neutrality tests, where the F_{ST} values of individual loci are tested against a simulated null distribution that is derived from a neutral island model of migration [137]. Another method widely used is the Bayesian method as implemented in BAYESCAN [138]. Alternative methods use environmental variables for outlier detection, implemented in software such as BAYENV2 [139] and BAYESCENV [140]. There are some constraints of F_{ST} outlier detection methods that should be considered, such as number of sampled demes [141] and population structure [142], and it does not display absolute differentiation, but differentiation relative to the total variation (see [143]). The menagerie of genomic tests for selection is the subject of a specialized literature (e.g., [144–146]) that is well beyond the scope of our review. In concert, these methods comprise a suite of tools, in some cases complementary, in others redundant. Overall, however, these methods are crucial to answering a question central to understanding the genomic architecture of differentiation and adaptive introgression in oaks and other species: which loci are exchanged between species, which are not, and under what spatial and environmental conditions does differentiation give way to introgression? Once identified, divergent loci can be localized in the genome using high resolution genetic linkage maps anchored to scaffolds of sequenced genomes [33,89,134].

Outlier loci have been identified for different oak species with contrasting ecological preferences, few of them showing very high interspecific differentiation as evidence for strong divergent selection. For example, one F_{ST} outlier (FIR013, an EST-SSR [147]) was identified when screening four *Q. rubra*/*Q. ellipsoidalis* species pairs from three different geographic regions using nuclear microsatellite markers and expressed sequence tag EST-SSR markers [82,136]. FIR013 was detected as an outlier in all geographic regions and pairwise comparisons, and was almost fixed for alternative alleles in the two species with values of interspecific F_{ST} ranging from 55% to 80%. The EST from which FIR013 was developed was annotated as *CONSTANS*-like (*COL*) gene, a candidate gene for flowering time and growth cessation in late summer [148]. *CONSTANS*-like genes are also involved in growth/development and phenology in other species [149,150], and it is thus possible that this gene may play a role in adaptive divergence between species through both ecological and phenological divergence [136]. Another outlier identified between *Q. rubra* and *Q. ellipsoidalis* included a histidine kinase 4-like gene (marker GOT021) [81], which was also detected as an outlier between *Q. pyrenaica* and *Q. faginea* [14]. Histidine kinase is shown to be an important part of a signaling cascade to effect stomatal closure in response to environmental and endogenous stimuli in *Arabidopsis* [151]. Additionally, GOT021 underlay a QTL for leaf shape variation in a *Q. robur* full-sib family [152]. The fact that it is an F_{ST} outlier in different sections suggests that this marker may be involved in the maintenance of species integrity across a broad phylogenetic range of oaks as a result of parallel environmental selection. The genomic architecture of such loci associated with species differentiation may be non-random and under parallel divergence pressure across the genus. A study of genomic coherence in eight species of the eastern North American white oak syngameon, for example, found SNPs that are fixed or nearly fixed across species' ranges [75]. SNPs divergent between species map back to all 12 *Quercus* linkage groups (chromosomes) and are separated from each other by an average of 7.47 million bp. The proportion of divergent SNPs separated by <10,000 bp is significantly higher than the null distribution, suggesting that genome-wide patterns of divergence may be concentrated on chromosomes or in regions of the genome that reflect a higher-than-average history of among-species divergence.

An outlier screen using genomic SSR markers widely distributed across the genome was performed in European white oaks as well [14,79,135]. Scotti-Saintagne et al. [79] recorded a potential hotspot of interspecific differentiation between *Q. robur* and *Q. petraea* on linkage group 12 (LG12). Goicoechea et al. [14] included related Mediterranean species, *Q. pyrenaica* and *Q. faginea*, in the analysis. Two genomic SSRs, QrZag87 (located on LG2) and QrZag112 (LG12), were identified as outliers in *Q. pyrenaica*/*Q. faginea* [14], as well as in *Q. robur*/*Q. petraea* [135].

Leroy et al. [44] performed genome-wide scans in *Q. petraea* populations sampled along latitudinal and elevational gradients to study local adaptation. Additionally, reference populations of other European white oak species were included (*Q. robur*, *Q. pubescens* and *Q. pyrenaica*) [47] to detect outliers potentially discovering footprints of divergence between these populations. To identify SNPs potentially subjected to selection within *Q. petraea*, XtX statistics (analogous to F_{ST} but specifically corrected for covariance matrix of allele frequency between populations, accounting for neutral correlations of allelic frequencies [139]) were calculated. 761,554 SNPs deviating from neutral expectations distributed over all chromosomes were detected, which is 2.05% of all investigated SNPs. Outliers were strongly enriched in SNPs which differentiated *Q. robur* from *Q. petraea*: strong correlations between intraspecific XtX values in 18 *Q. petraea* populations and interspecific F_{ST} between *Q. robur* and *Q. petraea* suggested that these SNPs are potentially under diversifying selection within and among species. Among the SNPs detected in *Q. petraea* association studies between genotype and environment/phenotype, 1331 SNPs were associated with temperature, 2932 with precipitation, and 1572 with leaf unfolding. The largest proportion of SNPs was significantly associated with both temperature and leaf unfolding, and those were highly enriched in SNPs strongly differentiated between *Q. robur* and *Q. petraea*. Subsequently, 167 unique candidate genes involved in different growth and developmental processes such as stomatal responses to water stress (e.g., *ALY4*, *ALMT9*, *PBL10*, *PP2AA*, *RopGEF1*, *SUS3*, *SCD1*) and some temperature associated genes acting as regulators of production of gibberellins were identified (e.g., *GAMT2*, *ASPG1* or *GA2ox*) [44].

In a companion paper, Leroy et al. [47] searched for candidate genes in regions enriched in outliers in the same four European white oak species based on a large data set of 31 million SNPs. In total, 227 genes were identified, from which at least 32 genes were assigned to three major functional categories (genes underlying ecological preferences of the four species, genes involved in biotic interactions, and those involved in intrinsic barriers (for the list of genes see Table 2 in [47])). Among these 32 genes, 6 candidate genes were detected as outliers for all species pairs. Four of these genes are putatively involved in intrinsic barriers, two are involved in pollen development (Cycloartenol synthase 2), and two in photoreceptor and UV-B tolerance. One of these—*UVR8*—was also detected as a gene under positive selection in *Juglans siligata* [153]. The other two outliers have a putative role in ecological barriers (regulator of root growth and role in soil-water deficit stress). Four outlier candidate genes related to drought tolerance were present in all observed species pairs except *Q. robur–Q. petraea* [47], which is consistent with the higher drought tolerance of *Q. pubescens* and *Q. pyrenaica* than in these two species [154]. One of the genes— *VRN1*, detected in all species pairs except in *Q. robur–Q. petraea*—plays a role in acclimation to low temperatures, and has been identified as a candidate gene for cold tolerance in many other plant species [155]. In general, most outlier SNPs contributing to reproductive barriers differentiate southern from northern species, suggesting that these barriers are driven by climate preferences [47]. Another association study on *Q. robur*, *Q. petraea*, and *Q. pubescens* was conducted by Rellstab et al. [156], who identified common patterns of SNP-environment associations across species for seven genes among 95 targeted genes, suggesting a role in local adaptation [156]. Genes *gigantea* and galactinol synthase (*GolS1*) from this study were also detected by Alberto et al. [157] as candidate genes for bud burst in *Q. petraea*. Additional genes showing patterns of local adaptation across different oak and related species of the beech family can be found in a recent review [158].

Whole genome outlier screens across species pairs can be used in combination with association analysis within species sampled across environmental gradients (Figure 2) to detect genes involved in both local adaptation within species and adaptive divergence between species. Such outliers, present in both analyses, are candidate genes for adaptive divergence and reproductive isolation between species [44,159]. Association analyses of genetic variation (single nucleotide polymorphisms, SNPs) with adaptive trait variation can be performed in segregating populations using QTL mapping approaches to detect associated genomic regions [79,134,160]. To narrow down these regions to individual genes, association mapping in unstructured populations such as provenance trials is required [133]. Additionally, environmental association analyses between SNP variation and environmental variables can be used to identify adaptive genetic variation [44,65,157,161]. Moreover, populations in extreme environments of the environmental cline can be selected for outlier analyses [33].

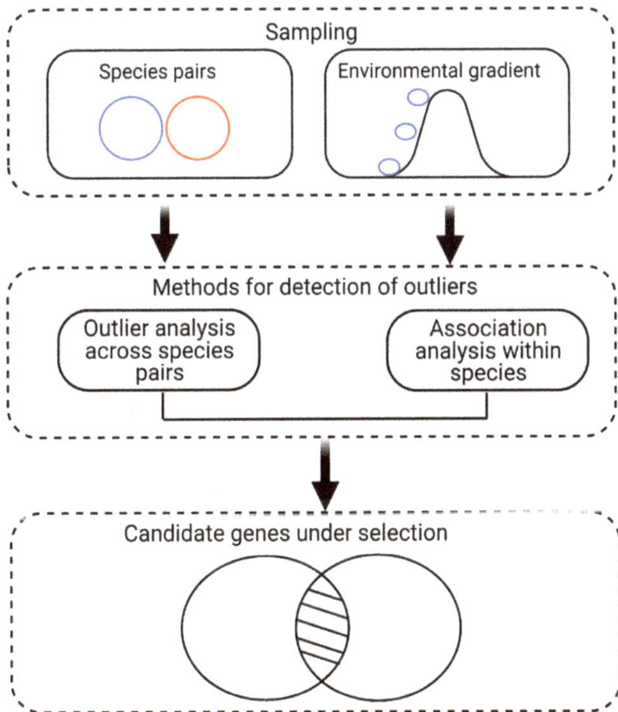

Figure 2. Workflow of detecting candidate genes under selection involved in adaptive divergence within and between species. Steps are explained in detail in the manuscript. Created with BioRender. com, accessed on 1 April 2021.

A final approach that has been illuminating in understanding the genomic architecture of differentiation in oaks is mapping phylogenetically informative loci back to the genome to determine whether (1) loci that recover introgression history are genomically clustered, following population-level studies of species divergence (e.g., [79]); or (2) loci that track divergence are replicated in different clades. Several studies have utilized phylogenomic datasets to explicitly model divergent vs. introgressive histories in oaks [69,90,99,162–167]. Two studies have then mapped loci implicated in ancient introgression vs. divergence in the white oaks (*Quercus* section *Quercus*) back to the genome to characterize the genomic distribution of divergence history [49,163]. Both studies failed to find that introgressive loci were clustered on the genome, presumably because the long history of recombination has eroded linkage disequilibrium connecting loci in ancient introgression events. Moreover, one of the studies [49] demonstrated that the loci recovering the divergence history in lieu of the dominant introgression history for alternative regions of the white phylogeny are also not related to one another. Thus, they reject the hypothesis that there are particular genes or regions of the genome that define the oak phylogeny globally. The evolutionary history of oaks is a mosaic history.

9. Conclusions and Future Perspectives

As summarized above, many studies have dealt with hybridization in oaks, but the decreasing costs and flexibility of whole genome sequencing and reduced representation genomic sequencing make it possible to study the evolutionary implications of hybridization and introgression at much higher genomic and spatial resolutions today [24]. To better understand the genetic basis for the maintenance of species integrity in oaks in the face of interspecific gene flow and the roles of genes that distinguish species, additional studies observing outlier regions between sympatric oak species should be performed. Analyzing oaks with different adaptations, for example to drought, might reveal whether there is a higher number of genes associated with adaptive divergence for drought tolerance and reproductive isolation between the ecologically divergent oak species in outlier genomic regions than in shared regions. Comparison of outlier genes identified between species with different environmental requirements across different sections can reveal genomic divergence unique to oak sections, or parallel genomic divergence driven by natural selection. Such outlier genes or regions detected across sections are prime candidates for adaptive species divergence by environmental selection. Whole genome resequencing can be used for the detection of such outliers and in combination with environmental association analysis, genome wide association studies (GWAS), and QTL mapping, candidate genes for speciation processes involved in adaptive divergence within and between species can be identified.

The focus of this review is solely on oaks and their maintenance of species integrity despite ongoing gene flow, hybridization, and introgression. Oaks, however, have the potential to illuminate speciation patterns in complement with immensely different speciation models such as sunflowers [168] and irises [169], where hybridization is a major driver of speciation. By contrast with the other models, in which hybridization triggers speciation [170], oaks exhibit a facilitation of adaptation via hybridization [44,48,69]. Equally important, hybridization appears not to threaten species integrity in oaks, even though a limited part of the genome—potentially a very limited portion, composed of only small regions distributed across the genome—is responsible for maintaining species barriers [24]. Thus, oaks are an alternative model of species differentiation, maintenance, and conservation that are equally important to understanding the tangled ways of speciation.

Author Contributions: Conceptualization, O.G., A.L.H., and J.E.C.; writing—original draft preparation, D.L.; writing—review and editing, O.G., A.L.H., J.E.C., and D.L.; supervision, O.G.; project administration, O.G.; funding acquisition, O.G. All authors have read and agreed to the published version of the manuscript.

Funding: This research was funded by Deutsche Forschungsgemeinschaft (DFG, German Research Foundation, Reference number: 429696097 (GA 714/7-1)).

Acknowledgments: We thank Katharina Birgit Budde for her support and constructive discussions.

Conflicts of Interest: The authors declare no conflict of interest. The funders had no role in the design of the study, the writing of the manuscript, or the decision to publish the results.

References

1. Darwin, C. *On the Origin of Species by Means of Natural Selection, or the Preservation of Favoured Races in the Struggle for Life*; John Murray: London, UK, 1859.
2. Schluter, D. Evidence for ecological speciation and its alternative. *Science* **2009**, *323*, 737–741. [CrossRef]
3. Via, S. Natural selection in action during speciation. *Proc. Natl. Acad. Sci. USA* **2009**, *106*, 9939–9946. [CrossRef] [PubMed]
4. Richards, E.J.; Servedio, M.R.; Martin, C.H. Searching for sympatric speciation in the genomic era. *BioEssays* **2019**, *41*. [CrossRef]
5. Matute, D.R.; Cooper, B.S. Comparative studies on speciation: 30 years since Coyne and Orr. *Evolution* **2021**, *75*, 764–778. [CrossRef] [PubMed]
6. Nosil, P.; Feder, J.L.; Gompert, Z. How many genetic changes create new species? *Science* **2021**, *371*, 777–780. [CrossRef] [PubMed]
7. Nosil, P. *Ecological Speciation*; Oxford University Press: Oxford, UK, 2012.
8. Nosil, P.; Feder, J.L. Genomic divergence during speciation: Causes and consequences. *Philos. Trans. R. Soc. B Biol. Sci.* **2012**, *367*, 332–342. [CrossRef]

9. Rundle, H.D.; Nosil, P. Ecological speciation. *Ecol. Lett.* **2005**, *8*, 336–352. [CrossRef]
10. Schluter, D. Ecology and the origin of species. *Trends Ecol. Evol.* **2001**, *16*, 372–380. [CrossRef]
11. Mayr, E. *Systematics and the Origin of Species*; Columbia University Press: New York, NY, USA, 1942.
12. Dobzhansky, T. *Genetics and the Origin of Species*; Columbia University Press: New York, NY, USA, 1937.
13. Coyne, J.A.; Orr, H.A. *Speciation*; Sinauer: Sunderland, MA, USA, 2004.
14. Goicoechea, P.G.; Herrán, A.; Durand, J.; Bodénès, C.; Plomion, C.; Kremer, A. A linkage disequilibrium perspective on the genetic mosaic of speciation in two hybridizing Mediterranean white oaks. *Heredity* **2015**, *114*, 373–386. [CrossRef]
15. Nosil, P.; Funk, D.J.; Ortiz-Barrientos, D. Divergent selection and heterogeneous genomic divergence. *Mol. Ecol.* **2009**, *18*, 375–402. [CrossRef]
16. Oleksyk, T.K.; Smith, M.W.; O'Brien, S.J. Genome-wide scans for footprints of natural selection. *Philos. Trans. R. Soc. B Biol. Sci.* **2010**, *365*, 185–205. [CrossRef]
17. Via, S.; West, J. The genetic mosaic suggests a new role for hitchhiking in ecological speciation. *Mol. Ecol.* **2008**, *17*, 4334–4345. [CrossRef]
18. Harr, B. Genomic islands of differentiation between house mouse subspecies. *Genome Res.* **2006**, *16*, 730–737. [CrossRef]
19. Yatabe, Y.; Kane, N.C.; Scotti-Saintagne, C.; Rieseberg, L.H. Rampant gene exchange across a strong reproductive barrier between the annual sunflowers, *Helianthus annuus* and *H. petiolaris*. *Genetics* **2007**, *175*, 1883–1893. [CrossRef] [PubMed]
20. Turner, T.L.; Hahn, M.W.; Nuzhdin, S.V. Genomic islands of speciation in *Anopheles gambiae*. *PLoS Biol.* **2005**, *3*, 1572–1578. [CrossRef]
21. Via, S. Divergence hitchhiking and the spread of genomic isolation during ecological speciation-with-gene-flow. *Philos. Trans. R. Soc. B Biol. Sci.* **2012**, *367*, 451–460. [CrossRef] [PubMed]
22. Brady, K.U.; Kruckeberg, A.R.; Bradshaw, H.D. Evolutionary ecology of plant adaptation to serpentine soils. *Annu. Rev. Ecol. Evol. Syst.* **2005**, *36*, 243–266. [CrossRef]
23. Lexer, C.; Lai, Z.; Rieseberg, L.H. Candidate gene polymorphisms associated with salt tolerance in wild sunflower hybrids: Implications for the origin of *Helianthus paradoxus*, a diploid hybrid species. *New Phytol.* **2004**, *161*, 225–233. [CrossRef]
24. Kremer, A.; Hipp, A.L. Oaks: An evolutionary success story. *New Phytol.* **2020**, *226*, 987–1011. [CrossRef]
25. Raeymaekers, J.A.M.; Chaturvedi, A.; Hablützel, P.I.; Verdonck, I.; Hellemans, B.; Maes, G.E.; De Meester, L.; Volckaert, F.A.M. Adaptive and non-adaptive divergence in a common landscape. *Nat. Commun.* **2017**, *8*, 1–8. [CrossRef] [PubMed]
26. Ågren, J.; Oakley, C.G.; Lundemo, S.; Schemske, D.W. Adaptive divergence in flowering time among natural populations of *Arabidopsis thaliana*: Estimates of selection and QTL mapping. *Evolution* **2017**, *71*, 550–564. [CrossRef]
27. Orr, H.A.; Masly, J.P.; Presgraves, D.C. Speciation genes. *Curr. Opin. Genet. Dev.* **2004**, *14*, 675–679. [CrossRef]
28. Wu, C.I.; Ting, C.T. Genes and speciation. *Nat. Rev. Genet.* **2004**, *5*, 114–122. [CrossRef] [PubMed]
29. Rieseberg, L.H.; Blackman, B.K. Speciation genes in plants. *Ann. Bot.* **2010**, *106*, 439–455. [CrossRef]
30. Nosil, P.; Schluter, D. The genes underlying the process of speciation. *Trends Ecol. Evol.* **2011**, *26*, 160–167. [CrossRef] [PubMed]
31. Leroy, T.; Plomion, C.; Kremer, A. Oak symbolism in the light of genomics. *New Phytol.* **2020**, *226*, 1012–1017. [CrossRef] [PubMed]
32. Neale, D.B.; Kremer, A. Forest tree genomics: Growing resources and applications. *Nat. Rev. Genet.* **2011**, *12*, 111–122. [CrossRef] [PubMed]
33. González-Martínez, S.C.; Krutovsky, K.V.; Neale, D.B. Forest-tree population genomics and adaptive evolution. *New Phytol.* **2006**, *170*, 227–238. [CrossRef] [PubMed]
34. Holliday, J.A.; Aitken, S.N.; Cooke, J.E.K.; Fady, B.; González-Martínez, S.C.; Heuertz, M.; Jaramillo-Correa, J.-P.; Lexer, C.; Staton, M.; Whetten, R.W.; et al. Advances in ecological genomics in forest trees and applications to genetic resources conservation and breeding. *Mol. Ecol.* **2017**, *26*, 706–717. [CrossRef]
35. Plomion, C.; Bastien, C.; Bogeat-Triboulot, M.B.; Bouffier, L.; Déjardin, A.; Duplessis, S.; Fady, B.; Heuertz, M.; Le Gac, A.L.; Le Provost, G.; et al. Forest tree genomics: 10 achievements from the past 10 years and future prospects. *Ann. For. Sci.* **2016**, *73*, 77–103. [CrossRef]
36. Tuscan, A.; DiFazio, S.; Jansson, S.; Putnam, N.; Ralph, S.; Rombauts, S.; Salamov, A.; Schein, J.; Sterck, L.; Aerts, A. The Genome of Black Cottonwood, *Populus trichocarpa* (Torr. & Gray). *Science* **2006**, *313*, 1596–1604.
37. Myburg, A.A.; Grattapaglia, D.; Tuskan, G.A.; Hellsten, U.; Hayes, R.D.; Grimwood, J.; Jenkins, J.; Lindquist, E.; Tice, H.; Bauer, D.; et al. The genome of *Eucalyptus grandis*. *Nature* **2014**, *510*, 356–362. [CrossRef]
38. Plomion, C.; Aury, J.M.; Amselem, J.; Leroy, T.; Murat, F.; Duplessis, S.; Faye, S.; Francillonne, N.; Labadie, K.; Le Provost, G.; et al. Oak genome reveals facets of long lifespan. *Nat. Plants* **2018**, *4*, 440–452. [CrossRef]
39. Nystedt, B.; Street, N.R.; Wetterbom, A.; Zuccolo, A.; Lin, Y.C.; Scofield, D.G.; Vezzi, F.; Delhomme, N.; Giacomello, S.; Alexeyenko, A.; et al. The Norway spruce genome sequence and conifer genome evolution. *Nature* **2013**, *497*, 579–584. [CrossRef]
40. Sork, V.L.; Fitz-Gibbon, S.T.; Puiu, D.; Crepeau, M.; Gugger, P.F.; Sherman, R.; Stevens, K.; Langley, C.H.; Pellegrini, M.; Salzberg, S.L. First draft assembly and annotation of the genome of a California endemic oak *Quercus lobata* Née (Fagaceae). *G3 Genes Genomes Genet.* **2016**, *6*, 3485–3495.
41. Ramos, A.M.; Usié, A.; Barbosa, P.; Barros, P.M.; Capote, T.; Chaves, I.; Simões, F.; Abreu, I.; Carrasquinho, I.; Faro, C.; et al. The draft genome sequence of cork oak. *Sci. Data* **2018**, *5*, 180069. [CrossRef] [PubMed]
42. Dodd, R.S.; Afzal-Rafii, Z. Selection and dispersal in a multispecies oak hybrid zone. *Soc. Study Evol.* **2004**, *58*, 261–269.

43. Howard, D.J.; Preszler, R.W.; Williams, J.; Fenchel, S.; Boecklen, W.J. How discrete are oak species? Insights from a hybrid zone between *Quercus grisea* and *Quercus gambelii*. *Evolution* **1997**, *51*, 747–755. [CrossRef]

44. Leroy, T.; Louvet, J.M.; Lalanne, C.; Le Provost, G.; Labadie, K.; Aury, J.M.; Delzon, S.; Plomion, C.; Kremer, A. Adaptive introgression as a driver of local adaptation to climate in European white oaks. *New Phytol.* **2020**, *226*, 1171–1182. [CrossRef]

45. Maze, J. Past Hybridization between *Quercus macrocarpa* and *Quercus gambelii*. *Brittonia* **1968**, *20*, 321–333. [CrossRef]

46. Muller, H.C. Ecological control of hybridization in *Quercus*: A factor in the mechanism of evolution. *Evolution* **1952**, *6*, 147–161.

47. Leroy, T.; Rougemont, Q.; Dupouey, J.L.; Bodénès, C.; Lalanne, C.; Belser, C.; Labadie, K.; Le Provost, G.; Aury, J.M.; Kremer, A.; et al. Massive postglacial gene flow between European white oaks uncovered genes underlying species barriers. *New Phytol.* **2019**. [CrossRef]

48. Leroy, T.; Roux, C.; Villate, L.; Bodénès, C.; Romiguier, J.; Paiva, J.A.P.; Dossat, C.; Aury, J.M.; Plomion, C.; Kremer, A. Extensive recent secondary contacts between four European white oak species. *New Phytol.* **2017**, *214*, 865–878. [CrossRef] [PubMed]

49. Hipp, A.L.; Manos, P.S.; Hahn, M.; Avishai, M.; Bodénès, C.; Cavender-Bares, J.; Crowl, A.A.; Deng, M.; Denk, T.; Fitz-Gibbon, S.; et al. Genomic landscape of the global oak phylogeny. *New Phytol.* **2020**, *226*, 1198–1212. [CrossRef] [PubMed]

50. Hubert, F.; Grimm, G.W.; Jousselin, E.; Berry, V.; Franc, A.; Kremer, A. Multiple nuclear genes stabilize the phylogenetic backbone of the genus *Quercus*. *Syst. Biodivers.* **2014**, *12*, 405–423. [CrossRef]

51. Manos, P.S.; Stanford, A.M. The historical biogeography of Fagaceae: Tracking the tertiary history of temperate and subtropical forests of the northern hemisphere. *Int. J. Plant Sci.* **2012**, *162*, S77–S93. [CrossRef]

52. Denk, T.; Grimm, G.W.; Manos, P.S.; Deng, M.; Hipp, A. An updated infrageneric classification of the oaks: Review of previous taxonomic schemes and synthesis of evolutionary patterns. In *Oaks Physiological Ecology. Exploring the Functional Diversity of Genus Quercus L.*; Springer International Publishing: New York, NY, USA, 2017; pp. 13–38.

53. Plomion, C.; Aury, J.M.; Amselem, J.; Alaeitabar, T.; Barbe, V.; Belser, C.; Berges, H.; Bodénès, C.; Boudet, N.; Boury, C.; et al. Decoding the oak genome: Public release of sequence data, assembly, annotation and publication strategies. *Mol. Ecol. Resour.* **2016**, *16*, 254–265. [CrossRef]

54. Sork, V.L.; Squire, K.; Gugger, P.F.; Steele, S.E.; Levy, E.D.; Eckert, A.J. Landscape genomic analysis of candidate genes for climate adaptation in a California endemic oak, *Quercus lobata*. *Am. J. Bot.* **2016**, *103*, 33–46. [CrossRef]

55. Aldrich, P.; Cavender-Bares, J. *Quercus*. In *Wild Crop Relatives: Genomic and Breeding Resources, Forest Trees*; Springer: Berlin, Germany, 2011; pp. 89–129.

56. Bar-Yosef, O.; Belfer-Cohen, A. Following Pleistocene road signs of human dispersals across Eurasia. *Quat. Int.* **2013**, *285*, 30–43. [CrossRef]

57. Cavender-Bares, J. Diversity, distribution and ecosystem services of the North American oaks. *Int. Oaks* **2016**, *27*, 37–48.

58. Chadwick, H.M. The Oak and the Thunder-God. *J. Anthropol. Inst. G. B. Irel.* **1900**, *30*, 22–44. [CrossRef]

59. Chassé, B. Eating Acorns: What Story Do the Distant, Far, and Near Past Tell Us, and Why? *Int. Oaks* **2016**, 107–135.

60. Logan, W.B. *Oak: The Frame of Civilization*; W.W. Norton & Company, Inc.: New York, NY, USA, 2005.

61. Thirgood, J. The historical significance of oak. In *Proceedings of the Oak Symposium*; U.S. Department of Agriculture, Forest Service, Northeastern Forest Experiment Station: Upper Darby, PA, USA; Morgantown, WV, USA, 1971.

62. Lev, E.; Kislev, M.E.; Bar-Yosef, O. Mousterian vegetal food in Kebara Cave, Mt. Carmel. *J. Archaeol. Sci.* **2004**, 475–484. [CrossRef]

63. Schmid-Siegert, E.; Sarkar, N.; Iseli, C.; Calderon, S.; Gouhier-Darimont, C.; Chrast, J.; Cattaneo, P.; Schütz, F.; Farinelli, L.; Pagni, M.; et al. Low number of fixed somatic mutations in a long-lived oak tree. *Nat. Plants* **2017**, *3*, 926–929. [CrossRef]

64. Rieseberg, L.H.; Wood, T.E.; Baack, E.J. The nature of plant species. *Nature* **2006**, *440*, 524–527. [CrossRef]

65. Reutimann, O.; Gugerli, F.; Rellstab, C. A species-discriminatory single-nucleotide polymorphism set reveals maintenance of species integrity in hybridizing European white oaks (*Quercus* spp.) despite high levels of admixture. *Ann. Bot.* **2020**, *125*, 663–676. [CrossRef]

66. Hipp, A.L.; Weber, J.A. Taxonomy of Hill's oak (*Quercus ellipsoidalis*: Fagaceae): Evidence from AFLP data. *Syst. Bot.* **2008**, *33*, 148–158. [CrossRef]

67. González-Rodríguez, A.; Oyama, K. Leaf morphometric variation in *Quercus affinis* and *Q. laurina* (Fagaceae), two hybridizing Mexican red oaks. *Bot. J. Linn. Soc.* **2005**, *147*, 427–435. [CrossRef]

68. González-Rodríguez, A.; Arias, D.M.; Valencia, S.; Oyama, K. Morphological and RAPD analysis of hybridization between *Quercus affinis* and *Q. laurina* (Fagaceae), two Mexican red oaks. *Am. J. Bot.* **2004**, *91*, 401–409. [CrossRef]

69. Eaton, D.A.R.; Hipp, A.L.; González-Rodríguez, A.; Cavender-Bares, J. Historical introgression among the American live oaks and the comparative nature of tests for introgression. *Evolution* **2015**, *69*, 2587–2601. [CrossRef] [PubMed]

70. Moran, E.V.; Willis, J.; Clark, J.S. Genetic evidence for hybridization in red oaks (*Quercus* sect. *Lobatae*, Fagaceae). *Am. J. Bot.* **2012**, *99*, 92–100. [CrossRef]

71. Van Valen, L. Ecological species, multispecies, and oaks. *Taxon* **1976**, *25*, 233–239. [CrossRef]

72. Burger, W.C. The species concept in *Quercus*. *Taxon* **1975**, *24*, 45–50. [CrossRef]

73. Cannon, C.H.; Petit, R.J. The oak syngameon: More than the sum of its parts. *New Phytol.* **2020**, *226*, 978–983. [CrossRef] [PubMed]

74. Cronk, Q.C.; Suarez-Gonzalez, A. The role of interspecific hybridization in adaptive potential at range margins. *Mol. Ecol.* **2018**, *27*, 4653–4656. [CrossRef] [PubMed]

75. Hipp, A.L.; Whittemore, A.T.; Garner, M.; Hahn, M.; Fitzek, E.; Guichoux, E.; Cavender-Bares, J.; Gugger, P.F.; Manos, P.S.; Pearse, I.S.; et al. Genomic identity of white oak species in an Eastern North American Syngameon. *Ann. Mo. Bot. Gard.* **2019**, *104*, 455–477. [CrossRef]

76. Lotsy, J. Species or Linneon. *Genetica* **1925**, *7*, 487–506. [CrossRef]

77. Khodwekar, S.; Gailing, O. Evidence for environment-dependent introgression of adaptive genes between two red oak species with different drought adaptations. *Am. J. Bot.* **2017**, *104*, 1088–1098. [CrossRef]

78. Sullivan, A.R.; Owusu, S.A.; Weber, J.A.; Hipp, A.L.; Gailing, O. Hybridization and divergence in multi-species oak (*Quercus*) communities. *Bot. J. Linn. Soc.* **2016**, *181*, 99–114. [CrossRef]

79. Scotti-Saintagne, C.; Mariette, S.; Porth, I.; Goicoechea, P.G.; Barreneche, T.; Bodénès, C.; Burg, K.; Kremer, A. Genome scanning for interspecific differentiation between two closely related oak species [*Quercus robur* L. and *Q. petraea* (Matt.) Liebl]. *Genetics* **2004**, *168*, 1615–1626. [CrossRef]

80. Curtu, A.L.; Gailing, O.; Finkeldey, R. Evidence for hybridization and introgression within a species-rich oak (*Quercus* spp.) community. *BMC Evol. Biol.* **2007**, *7*, 218. [CrossRef] [PubMed]

81. Lind, J.F.; Gailing, O. Genetic structure of *Quercus rubra* L. and *Quercus ellipsoidalis* E. J. Hill populations at gene-based EST-SSR and nuclear SSR markers. *Tree Genet. Genomes* **2013**, *9*, 707–722. [CrossRef]

82. Collins, E.; Sullivan, A.R.; Gailing, O. Limited effective gene flow between two interfertile red oak species. *Trees Struct. Funct.* **2015**, *29*, 1135–1148. [CrossRef]

83. Owusu, S.A.; Sullivan, A.R.; Weber, J.A.; Hipp, A.L.; Gailing, O. Taxonomic relationships and gene flow in four North American *Quercus* species (*Quercus* section *Lobatae*). *Syst. Bot.* **2015**, *40*, 510–521. [CrossRef]

84. Cavender-Bares, J.; Ackerly, D.D.; Baum, D.A.; Bazzaz, F.A. Phylogenetic overdispersion in Floridian oak communities. *Am. Nat.* **2004**, *163*, 823–843. [CrossRef]

85. Gailing, O. Differences in growth, survival and phenology in *Quercus rubra* and *Q. ellipsoidalis* seedlings. *Dendrobiology* **2013**, *70*, 73–81. [CrossRef]

86. López De Heredia, U.; Valbuena-Carabaña, M.; Córdoba, M.; Gil, L. Variation components in leaf morphology of recruits of two hybridising oaks [*Q. petraea* (Matt.) Liebl. and *Q. pyrenaica* Willd.] at small spatial scale. *Eur. J. For. Res.* **2009**, *128*, 543–554. [CrossRef]

87. Valbuena-Carabaña, M.; González-Martínez, S.C.; Sork, V.L.; Collada, C.; Soto, A.; Goicoechea, P.G.; Gil, L. Gene flow and hybridisation in a mixed oak forest (*Quercus pyrenaica* Willd. and *Quercus petraea* (Matts.) Liebl.) in central Spain. *Heredity* **2005**, *95*, 457–465. [CrossRef]

88. Cottam, W.; Tucker, J.; Santamour, F. *Oak Hybridization at the University of Utah*; State Arboretum of Utah: Salt Lake City, UT, USA, 1982.

89. Gailing, O.; Curtu, A.L. Interspecific gene flow and maintenance of species integrity in oaks. *Ann. For. Res.* **2014**, *57*, 5–18. [CrossRef]

90. McVay, J.D.; Hipp, A.L.; Manos, P.S. A genetic legacy of introgression confounds phylogeny and biogeography in oaks. *Proc. R. Soc. B Biol. Sci.* **2017**, *284*, 20170300. [CrossRef]

91. Manos, P.S.; Doyle, J.J.; Nixon, K.C. Phylogeny, biogeography, and processes of molecular differentiation in *Quercus* subgenus *Quercus* (Fagaceae). *Mol. Phylogenet. Evol.* **1999**, *12*, 333–349. [CrossRef]

92. Pham, K.K.; Hipp, A.L.; Manos, P.S.; Cronn, R.C. A time and a place for everything: Phylogenetic history and geography as joint predictors of oak plastome phylogeny. *Genome* **2017**, *60*, 720–732. [CrossRef]

93. Abrams, M.D. Comparative water relations of three successional hardwood species in central Wisconsin. *Tree Physiol.* **1988**, *4*, 263–273. [CrossRef] [PubMed]

94. Abrams, M.D. Adaptations and responses to drought in *Quercus* species of North America. *Tree Physiol.* **1990**, *7*, 227–238. [CrossRef] [PubMed]

95. Bréda, N.; Cochard, H.; Dreyer, E.; Granier, A. Field comparison of transpiration, stomatal conductance and vulnerability to cavitation of *Quercus petraea* and *Quercus robur* under water stress. *Ann. des Sci. For.* **1993**, *50*, 571–582. [CrossRef]

96. Cavender-Bares, J.; Pahlich, A. Molecular, morphological, and ecological niche differentiation of sympatric sister oak species, *Quercus virginiana* and *Q. geminata* (Fagaceae). *Am. J. Bot.* **2009**, *96*, 1690–1702. [CrossRef] [PubMed]

97. Petit, R.J.; Bodénès, C.; Ducousso, A.; Roussel, G.; Kremer, A. Hybridization as a mechanism of invasion in oaks. *New Phytol.* **2003**, *161*, 151–164. [CrossRef]

98. Curtu, A.L.; Gailing, O.; Finkeldey, R. Patterns of contemporary hybridization inferred from paternity analysis in a four-oak-species forest. *BMC Evol. Biol.* **2009**, *9*, 284. [CrossRef]

99. McVay, J.D.; Hauser, D.; Hipp, A.L.; Manos, P.S. Phylogenomics reveals a complex evolutionary history of lobed-leaf white oaks in western North America. *Genome* **2017**, *60*, 733–742. [CrossRef]

100. Lepais, O.; Petit, R.J.; Guichoux, E.; Lavabre, J.E.; Alberto, F.; Kremer, A.; Gerber, S. Species relative abundance and direction of introgression in oaks. *Mol. Ecol.* **2009**, *18*, 2228–2242. [CrossRef]

101. Anderson, R.; Harrison, T. A limitation of the hybrid index using *Quercus* leaf characters. *Southwest. Nat.* **1979**, *24*, 463–473. [CrossRef]

102. Bacon, J.; Spellenberg, R. Hybridization in two distantly related Mexican black oaks *Quercus conzattii* and *Quercus eduardii* (Fagaceae: *Quercus*: Section *Lobatae*). *SIDA Contrib. Bot.* **1996**, *17*, 17–41.

103. Bartlett, H. Regression of x *Quercus deamii* toward *Quercus macrocarpa* and *Quercus muhlenbergii*. *Rhodora* **1951**, *53*, 249–264.
104. Jensen, R.J.; DePiero, R.; Smith, B. Vegetative characters, population variation and the hybrid origin of *Quercus ellipsoidalis*. *Am. Midl. Nat.* **1984**, *111*, 364–370. [CrossRef]
105. Jensen, J. A preliminary numerical analysis of the red oak complex in Michigan and Wisconsin. *Taxon* **1997**, *26*, 399–407. [CrossRef]
106. Rushton, B. Natural hybridization within the genus *Quercus* L. *Ann. Sci. For.* **1993**, *50*, 73s–90s. [CrossRef]
107. Ortego, J.; Gugger, P.F.; Riordan, E.C.; Sork, V.L. Influence of climatic niche suitability and geographical overlap on hybridization patterns among southern Californian oaks. *J. Biogeogr.* **2014**, *41*, 1895–1908. [CrossRef]
108. Craft, K.J.; Ashley, M.V.; Koenig, W.D. Limited hybridization between *Quercus lobata* and *Quercus douglasii* (Fagaceae) in a mixed stand in central coastal California. *Am. J. Bot.* **2002**, *89*, 1792–1798. [CrossRef] [PubMed]
109. Ramos-Ortiz, S.; Oyama, K.; Rodríguez-Correa, H.; González-Rodríguez, A. Geographic structure of genetic and phenotypic variation in the hybrid zone between *Quercus affinis* and *Q. laurina* in Mexico. *Plant Species Biol.* **2016**, *31*, 219–232. [CrossRef]
110. Rellstab, C.; Bühler, A.; Graf, R.; Folly, C.; Gugerli, F. Using joint multivariate analyses of leaf morphology and molecular-genetic markers for taxon identification in three hybridizing European white oak species (*Quercus* spp.). *Ann. For. Sci.* **2016**, *73*, 669–679. [CrossRef]
111. Yücedağ, C.; Müller, M.; Gailing, O. Morphological and genetic variation in natural populations of *Quercus vulcanica* and *Q. frainetto*. *Plant Syst. Evol.* **2021**, *307*, 8. [CrossRef]
112. Fitzek, E.; Delcamp, A.; Guichoux, E.; Hahn, M.; Lobdell, M.; Hipp, A.L. A nuclear DNA barcode for eastern North American oaks and application to a study of hybridization in an Arboretum setting. *Ecol. Evol.* **2018**, *8*, 5837–5851. [CrossRef]
113. Guichoux, E.; Lagache, L.; Wagner, S.; Léger, P.; Petit, R.J. Two highly validated multiplexes (12-plex and 8-plex) for species delimitation and parentage analysis in oaks (*Quercus* spp.). *Mol. Ecol. Resour.* **2011**, *11*, 578–585. [CrossRef] [PubMed]
114. Lepais, O.; Gerber, S. Reproductive patterns shape introgression dynamics and species succession within the European white oak species complex. *Evolution* **2011**, *65*, 156–170. [CrossRef] [PubMed]
115. Salvini, D.; Bruschi, P.; Fineschi, S.; Grossoni, P.; Kjaer, E.D.; Vendramin, G.G. Natural hybridisation between *Quercus petraea* (Matt.) Liebl. and *Quercus pubescens* Willd. within an Italian stand as revealed by microsatellite fingerprinting. *Plant Biol.* **2009**, *11*, 758–765. [CrossRef]
116. Pritchard, J.K.; Stephens, M.; Donnelly, P. Inference of population structure using multilocus genotype data. *Genetics* **2000**, *155*, 945–959. [CrossRef]
117. Abraham, S.T.; Zaya, D.N.; Koenig, W.D.; Ashley, M.V. Interspecific and intraspecific pollination patterns of valley oak, *Quercus lobata*, in a mixed stand in coastal central California. *Int. J. Plant Sci.* **2011**, *172*, 691–699. [CrossRef]
118. Zhang, R.; Hipp, A.L.; Gailing, O. Sharing of chloroplast haplotypes among red oak species suggests interspecific gene flow between neighboring populations. *Botany* **2015**, *93*, 691–700. [CrossRef]
119. Muir, G.; Fleming, C.C.; Schlötterer, C. Species status of hybridizing oaks. *Nature* **2000**, *405*, 1016. [CrossRef]
120. Aldrich, P.R.; Parker, G.R.; Michler, C.H.; Romero-Severson, J. Whole-tree silvic identifications and the microsatellite genetic structure of a red oak species complex in an Indiana old-growth forest. *Can. J. For. Res.* **2003**, *33*, 2228–2237. [CrossRef]
121. Craft, K.J.; Ashley, M.V. Population differentiation among three species of white oak in northeastern Illinois. *Can. J. For. Res.* **2006**, *36*, 206–215. [CrossRef]
122. Gailing, O.; Zhang, R. Experimental evidence for selection against hybrids between two interfertile red oak species. *Silvae Genet.* **2019**, *67*, 106–110. [CrossRef]
123. Anderson, E. *Introgressive Hybridization*; John Wiley Sons, Inc.: New York, NY, USA, 1949.
124. Anderson, E. Hybridization of the habitat. *Evolution* **1948**, *2*, 1–9. [CrossRef]
125. Cavender-Bares, J.; Keen, A.; Miles, B. Phylogenetic structure of Floridian plant communities depends on taxonomic and spatial scale. *Ecology* **2006**, *87*, 109–122. [CrossRef]
126. Hipp, A.L.; Manos, P.S.; González-Rodríguez, A.; Hahn, M.; Kaproth, M.; McVay, J.D.; Avalos, S.V.; Cavender-Bares, J. Sympatric parallel diversification of major oak clades in the Americas and the origins of Mexican species diversity. *New Phytol.* **2018**, *217*, 439–452. [CrossRef]
127. Jensen, J.; Larsen, A.; Nielsen, L.R.; Cottrell, J. Hybridization between *Quercus robur* and *Q. petraea* in a mixed oak stand in Denmark. *Ann. For. Sci.* **2009**, *66*, 706. [CrossRef]
128. Zeng, Y.F.; Liao, W.J.; Petit, R.J.; Zhang, D.Y. Geographic variation in the structure of oak hybrid zones provides insights into the dynamics of speciation. *Mol. Ecol.* **2011**, *20*, 4995–5011. [CrossRef] [PubMed]
129. Lind-Riehl, J.F.; Gailing, O. Adaptive variation and introgression of a CONSTANS-like gene in North American red oaks. *Forests* **2017**, *8*, 3. [CrossRef]
130. Suarez-Gonzalez, A.; Lexer, C.; Cronk, Q.C.B. Adaptive introgression: A plant perspective. *Biol. Lett.* **2018**, *14*, 20170688. [CrossRef] [PubMed]
131. Li, Z.; Liu, D. ROPGEF1 and ROPGEF4 are functional regulators of ROP11 GTPase in ABA-mediated stomatal closure in *Arabidopsis*. *FEBS Lett.* **2012**, *586*, 1253–1258. [CrossRef]
132. Elhaddad, N.S.; Hunt, L.; Sloan, J.; Gray, J.E. Light-induced stomatal opening is affected by the guard cell protein kinase APK1b. *PLoS ONE* **2014**, *9*, e97161. [CrossRef] [PubMed]
133. Browne, L.; Wright, J.W.; Fitz-Gibbon, S.; Gugger, P.F.; Sork, V.L. Adaptational lag to temperature in valley oak (*Quercus lobata*) can be mitigated by genome-informed assisted gene flow. *Proc. Natl. Acad. Sci. USA* **2019**, *116*, 25179–25185. [CrossRef] [PubMed]

134. Bodénès, C.; Chancerel, E.; Ehrenmann, F.; Kremer, A.; Plomion, C. High-density linkage mapping and distribution of segregation distortion regions in the oak genome. *DNA Res.* **2016**, *23*, 115–124. [CrossRef] [PubMed]

135. Goicoechea, P.G.; Petit, R.J.; Kremer, A. Detecting the footprints of divergent selection in oaks with linked markers. *Heredity* **2012**, *109*, 361–371. [CrossRef] [PubMed]

136. Lind-Riehl, J.F.; Sullivan, A.R.; Gailing, O. Evidence for selection on a CONSTANS-like gene between two red oak species. *Ann. Bot.* **2014**, *113*, 967–975. [CrossRef] [PubMed]

137. Beaumont, M.A.; Nichols, R.A. Evaluating loci for use in the genetic analysis of population structure. *Proc. R. Soc. B Biol. Sci.* **1996**, *263*, 1619–1626.

138. Foll, M.; Gaggiotti, O. A genome-scan method to identify selected loci appropriate for both dominant and codominant markers: A Bayesian perspective. *Genetics* **2008**, *180*, 977–993. [CrossRef]

139. Günther, T.; Coop, G. Robust identification of local adaptation from allele frequencies. *Genetics* **2013**, *195*, 205–220. [CrossRef]

140. de Villemereuil, P.; Gaggiotti, O.E. A new F_{ST}-based method to uncover local adaptation using environmental variables. *Methods Ecol. Evol.* **2015**, *6*, 1248–1258. [CrossRef]

141. Caballero, A.; Quesada, H.; Rolán-Alvarez, E. Impact of amplified fragment length polymorphism size homoplasy on the estimation of population genetic diversity and the detection of selective loci. *Genetics* **2008**, *179*, 539–554. [CrossRef]

142. Excoffier, L.; Hofer, T.; Foll, M. Detecting loci under selection in a hierarchically structured population. *Heredity* **2009**, *103*, 285–298. [CrossRef] [PubMed]

143. Flanagan, S.P.; Jones, A.G. Constraints on the F_{ST}—Heterozygosity outlier approach. *J. Hered.* **2017**, *108*, 561–573. [CrossRef] [PubMed]

144. Forester, B.R.; Lasky, J.R.; Wagner, H.H.; Urban, D.L. Comparing methods for detecting multilocus adaptation with multivariate genotype-environment associations. *Mol. Ecol.* **2018**, *27*, 2215–2233. [CrossRef]

145. Galesloot, T.E.; Van Steen, K.; Kiemeney, L.A.L.M.; Janss, L.L.; Vermeulen, S.H. A comparison of multivariate genome-wide association methods. *PLoS ONE* **2014**, *9*, e95923. [CrossRef]

146. Martins, H.; Caye, K.; Luu, K.; Blum, M.G.B.; François, O. Identifying outlier loci in admixed and in continuous populations using ancestral population differentiation statistics. *Mol. Ecol.* **2016**, *25*, 5029–5042. [CrossRef] [PubMed]

147. Durand, J.; Bodénès, C.; Chancerel, E.; Frigerio, J.M.; Vendramin, G.; Sebastiani, F.; Buonamici, A.; Gailing, O.; Koelewijn, H.P.; Villani, F.; et al. A fast and cost-effective approach to develop and map EST-SSR markers: Oak as a case study. *BMC Genom.* **2010**, *11*, 570. [CrossRef] [PubMed]

148. Ano, M.; Katayose, Y.; Ashikari, M.; Yamanouchi, U.; Monna, L.; Fuse, T.; Baba, T.; Yamamoto, K.; Umehara, Y.; Nagamura, Y.; et al. Hd1, a major photoperiod sensitivity quantitative trait locus in rice, is closely related to the *Arabidopsis* flowering time gene CONSTANS. *Plant. Cell* **2000**, *12*, 2473–2483. [CrossRef]

149. Herrmann, D.; Barre, P.; Santoni, S.; Julier, B. Association of a CONSTANS-LIKE gene to flowering and height in autotetraploid alfalfa. *Theor. Appl. Genet.* **2010**, *121*, 865–876. [CrossRef]

150. Hsu, C.Y.; Adams, J.P.; No, K.; Liang, H.; Meilan, R.; Pechánová, O.; Barakat, A.; Carlson, J.E.; Page, G.P.; Yuceer, C. Overexpression of constans homologs CO1 and CO2 fails to alter normal reproductive onset and fall bud set in woody perennial poplar. *PLoS ONE* **2012**, *7*, e45448. [CrossRef]

151. Desikan, R.; Horák, J.; Chaban, C.; Mira-Rodado, V.; Witthöft, J.; Elgass, K.; Grefen, C.; Cheung, M.K.; Meixner, A.J.; Hooley, R.; et al. The histidine kinase AHK5 integrates endogenous and environmental signals in *Arabidopsis* guard cells. *PLoS ONE* **2008**, *3*, e2491. [CrossRef]

152. Gailing, O.; Bodénès, C.; Finkeldey, R.; Kremer, A.; Plomion, C. Genetic mapping of EST-derived simple sequence repeats (EST-SSRs) to identify QTL for leaf morphological characters in a *Quercus robur* full-sib family. *Tree Genet. Genomes* **2013**, *9*, 1361–1367. [CrossRef]

153. Ning, D.L.; Wu, T.; Xiao, L.J.; Ma, T.; Fang, W.L.; Dong, R.Q.; Cao, F.L. Chromosomal-level assembly of *Juglans sigillata* genome using Nanopore, BioNano, and Hi-C analysis. *Gigascience* **2020**, *9*, giaa006. [CrossRef] [PubMed]

154. Fonti, P.; Heller, O.; Cherubini, P.; Rigling, A.; Arend, M. Wood anatomical responses of oak saplings exposed to air warming and soil drought. *Plant. Biol.* **2013**, *15*, 210–219. [CrossRef]

155. Levy, Y.Y.; Mesnage, S.; Mylne, J.S.; Gendall, A.R.; Dean, C. Multiple roles of *Arabidopsis* VRN1 in vernalization and flowering time control. *Science* **2002**, *297*, 243–246. [CrossRef] [PubMed]

156. Rellstab, C.; Zoller, S.; Walthert, L.; Lesur, I.; Pluess, A.R.; Graf, R.; Bodénès, C.; Sperisen, C.; Sperisen, C.; Kremer, A.; et al. Signatures of local adaptation in candidate genes of oaks (*Quercus* spp.) with respect to present and future climatic conditions. *Mol. Ecol.* **2016**, *25*, 5907–5924. [CrossRef] [PubMed]

157. Alberto, F.J.; Derory, J.; Boury, C.; Frigerio, J.-M.; Zimmermann, N.E.; Kremer, A. Imprints of natural selection along environmental gradients in phenology-related genes of *Quercus petraea*. *Genetics* **2013**, *195*, 495–512. [CrossRef]

158. Müller, M.; Gailing, O. Abiotic genetic adaptation in the Fagaceae. *Plant. Biol.* **2019**, *21*, 783–795. [CrossRef]

159. Gailing, O. Identification of genes under divergent selection in interfertile, but ecologically divergent oaks. *Bull. Transilv. Univ. Bras. Ser. II For. Wood Ind. Agric. Food Eng.* **2014**, *7*, 1–10.

160. López de Heredia, U.; Mora-Márquez, F.; Goicoechea, P.G.; Guillardín-Calvo, L.; Simeone, M.C.; Soto, A. ddRAD sequencing-based identification of genomic boundaries and permeability in *Quercus ilex* and *Q. suber* hybrids. *Front. Plant. Sci.* **2020**, *11*, 564414. [CrossRef]

161. Gugger, P.F.; Fitz-Gibbon, S.T.; Albarrán-Lara, A.; Wright, J.W.; Sork, V.L. Landscape genomics of *Quercus lobata* reveals genes involved in local climate adaptation at multiple spatial scales. *Mol. Ecol.* **2021**, *30*, 406–423. [CrossRef]
162. Cavender-Bares, J.; González-Rodríguez, A.; Eaton, D.A.R.; Hipp, A.A.L.; Beulke, A.; Manos, P.S. Phylogeny and biogeography of the American live oaks (*Quercus* subsection *Virentes*): A genomic and population genetics approach. *Mol. Ecol.* **2015**, *24*, 3668–3687. [CrossRef] [PubMed]
163. Crowl, A.A.; Manos, P.S.; McVay, J.D.; Lemmon, A.R.; Lemmon, E.M.; Hipp, A.L. Uncovering the genomic signature of ancient introgression between white oak lineages (*Quercus*). *New Phytol.* **2020**, *226*, 1158–1170. [CrossRef] [PubMed]
164. Crowl, A.; Bruno, E.; Hipp, A.; Manos, P. Revisiting the mystery of the Bartram oak. *Arnoldia* **2020**, *77*, 6–11.
165. Hauser, D.A.; Keuter, A.; McVay, J.D.; Hipp, A.L.; Manos, P.S. The evolution and diversification of the red oaks of the California floristic province (*Quercus* section *Lobatae*, series *Agrifoliae*). *Am. J. Bot.* **2017**, *104*, 1581–1595. [CrossRef]
166. Kim, B.Y.; Wei, X.; Fitz-Gibbon, S.; Lohmueller, K.E.; Ortego, J.; Gugger, P.F.; Sork, V.L. RADseq data reveal ancient, but not pervasive, introgression between Californian tree and scrub oak species (*Quercus* sect. *Quercus*: Fagaceae). *Mol. Ecol.* **2018**, *27*, 4556–4571. [CrossRef] [PubMed]
167. Ortego, J.; Gugger, P.F.; Sork, V.L. Genomic data reveal cryptic lineage diversification and introgression in Californian golden cup oaks (section *Protobalanus*). *New Phytol.* **2018**, *218*, 804–818. [CrossRef] [PubMed]
168. Rieseberg, L.H. Hybrid Speciation in Wild Sunflowers. *Ann. Mo. Bot. Gard.* **2006**, *93*, 34–48. [CrossRef]
169. Arnold, M.L.; Bouck, A.C.; Cornman, R.S. Verne Grant and Louisiana irises: Is there anything new under the sun? *New Phytol.* **2004**, *161*, 143–149. [CrossRef]
170. Abbott, R.; Albach, D.; Ansell, S.; Arntzen, J.W.; Baird, S.J.E.; Bierne, N.; Boughman, J.; Brelsford, A.; Buerkle, C.A.; Buggs, R.; et al. Hybridization and speciation. *J. Evol. Biol.* **2013**, *26*, 229–246. [CrossRef] [PubMed]

Viewpoint

Answers Blowing in the Wind: A Quarter Century of Genetic Studies of Pollination in Oaks

Mary V. Ashley

Department of Biological Sciences, University of Illinois at Chicago, Chicago, IL 60607, USA; ashley@uic.edu

Abstract: For the past 25 years, the twin tools of highly variable genetic markers (microsatellites) and paternity assignment have provided a powerful approach for investigating pollination patterns in trees, including many *Quercus* species. Early studies consistently demonstrated surprisingly abundant and extensive long-distance pollen movement in oaks. Indeed, numerous studies showed high levels of pollen immigration (50% or more), even for relatively isolated stands of oaks. Research also characterized fertilization patterns within stands and between hybridizing species in mixed stands. More recent studies have expanded our knowledge of genetic exchange effected by successful pollen movement, identified even more remarkable examples of the distances *Quercus* pollen can travel, and examined pollination patterns in relictual populations as well as those at the leading edges of range expansion. While the paradigm of long distance pollination continues to hold, a few recent studies that have also revealed the limits of pollen movement, identifying cases of reproductive isolation in extreme situations, where populations are at risk. This review will highlight what has been learned about *Quercus* pollination, what questions remain, and propose implications for forest management in the face of changing landscapes and climates.

Keywords: *Quercus*; pollination; microsatellites; parentage analysis; gene flow; wind pollination

Citation: Ashley, M.V. Answers Blowing in the Wind: A Quarter Century of Genetic Studies of Pollination in Oaks. *Forests* **2021**, *12*, 575. https://doi.org/10.3390/f12050575

Academic Editor: Stefan Arndt

Received: 19 April 2021
Accepted: 2 May 2021
Published: 5 May 2021

Publisher's Note: MDPI stays neutral with regard to jurisdictional claims in published maps and institutional affiliations.

Tracking pollination patterns is critical for understanding many fundamental aspects of *Quercus* biology, including reproductive patterns, impacts of habitat fragmentation, responses to climate change, and even masting dynamics. Prior to the availability of genetic markers to track pollen movement, characterizing pollination patterns in plants was difficult, especially for plants with wind-dispersed pollen such as members of the genus *Quercus*. Indirect approaches using pollen traps or dyes are logistically challenging, and only track the movement of pollen, not actual fertilizations. It has been about a quarter of a century since a class of genetic markers known as microsatellites were first applied to questions of pollen and seed dispersal in trees [1–4]. Also known as simple sequence repeats (SSRs) or short tandem repeats (STRs), these DNA sequences are repeated a dozen or more times. An important characteristic of microsatellites, and the source of their utility, is a high mutation rate for their number of repeat units. As a result, each microsatellite locus has many "alleles," variants that differ only by their number of repeats. These loci can be scored for length variants using a polymerase chain reaction (PCR) assay, and they provide a rich source of genetic markers for mating system studies such as the assignment of paternity [1,5].

Oaks were among the first systems where microsatellites combined with paternity assignment were used for describing pollination patterns [3,4,6]. Oaks are diploid, wind-pollinated and tend to occur in low diversity forests. The general approach for such studies would be to select an isolated stand or a portion of one where the density of trees was relatively low, so that a manageable number of reproductive adult trees could all be sampled. These adult trees would be genotyped at several microsatellite loci (typically 6–10). Acorns (or seedlings grown from acorns) would be sampled from a set of "maternal trees" and also genotyped. At every locus, the seed would have one maternal allele and one paternal allele. The genotypes of the other adults in the stand would be compared to assign paternity to one tree that could have contributed all of the paternal alleles. If no adult tree can be assigned, this indicated that the paternal tree was located beyond the study site and pollen immigration had occurred.

The concept of paternity assignment to track pollination patterns is straightforward, but in practice, there are several challenges that can complicate these studies. Advances in genetic marker availability and genotyping have helped ease technical issues that can occur in the lab, but genotyping is still not foolproof, and a small rate of error (including null alleles) must be accounted for. Paternity assignment by exclusion is unambiguous in principle but quite thorny in practice. Likelihood methods are usually employed but require estimates of parameters that may be uncertain, such as the proportion of candidate fathers sampled. Many of these issues and limitations are considered in more detail in Ashley [5]. For purposes of this review, I include studies conducted with an adequate set of markers and established analytical approaches, and I report the authors' conclusions regarding observed pollination patterns.

After developing the first microsatellite markers for oaks [7], Dow and Ashley investigated pollination and seed dispersal in a stand of bur oaks, *Quercus macrocarpa*, in Northern Illinois, USA [3,4,8]. Although the stand of 62 adult trees was relatively isolated, the researchers discovered that the majority of pollinations (62%) were affected by trees outside the stand, at minimal distances of over 150 m. For within-stand pollinations, the average pollination distance was 60 m. Pollen donors were located in all parts of the stand and had a nearly random distribution around the maternal trees. These findings countered the prevailing view of pollination at the time, which envisioned a leptokurtic distribution around a single source with concentrations of pollen dropping rapidly with distance [9,10]. Indeed, distance between trees in a stand seemed to explain very little about the observed pollination patterns. Further investigations on *Q. macrocarpa* demonstrated that even extremely isolated stands of oaks receive large amounts of pollen from distant sources [11], and as a result, may be quite resilient to negative genetic consequences of habitat fragmentation [12].

Reports of substantial levels of long distance pollination in other species of *Quercus* in different habitats and continents began to appear. In a mixed stand of the European species *Quercus robur* and *Q. petraea*, gene flow from outside the stand exceeded 65% for both species [6]. In a later study involving a mixed stand of *Q. petraea* and *Q. pyrenaica* in central Spain [13], over a third of pollinations came from outside the stand for both species. Remarkably, within stand pollinations for *Q. pyrenaica* averaged 270 m, comparable to the average pairwise distance between trees. In Japan, studies of *Q. salicina* showed a pollen immigration rate of 52% [14,15]. A study of *Quercus lobata* in Coastal Central California found that 70% of effective pollen traveled more than 200 m [16]. Thus, about a decade ago, long distance pollination and high rates of pollen immigration emerged as the new paradigm for *Quercus* (and other wind pollinated trees), as a result of the application of paternity assignment and the availability of highly variable microsatellite markers [5].

These studies reported high levels of gene flow and pollination distances exceeding 100 m, but they were constrained by sampling limitations in detecting just how far *Quercus* pollen might travel. The maximum distance that can be detected is the maximum distance between sampled trees in the study plot. Buschbom et al. [17] overcame that limitation by sampling a relict stand of *Q. robur* east of the Ural Mountains near Sibay, Russia. The stand is 30 km from another small relic stand and at least 80 km from the next closest oak occurrences. Remarkably, at least 35% of the acorns sampled there were the result of long-distance pollination, and most could not be assigned to trees in the other relict stand. This pushed the record of *Quercus* pollination distances to tens of kilometers. Substantial levels of pollen immigration have been reported at other isolated sites of *Quercus* (although none as isolated as the Sibay *Q. robur* stand). These include relict stands of *Q. macrocarpa* in agricultural landscapes of the midwestern USA [11] and low-density and highly fragmented stands of *Q. ilex* in Central Spain [18].

Studies began to investigate pollen movement in the context of deforestation and anthropogenic landscape changes. One interesting study was conducted on *Quercus castanea*, a species distributed from northern Mexico to Guatemala [19]. At the study site in Central Mexico, *Q. castanea* grew in fragmented forest patches and as isolated remnant trees. Gene flow via pollen was not reduced for the isolated trees, and the authors conclude that these scattered surviving trees facilitate gene flow and connectivity among forest patches. Another recent study of fragmented oak forests was conducted on *Q. bambusifolia* in Hong Kong [20]. While this study did not track contemporary pollen movement using paternity assignment, it did provide evidence for the role of pollen movement in the genetic recovery of *Q. bambusifolia* on Hong Kong Island since WWII, when forest cover was reduced to nearly zero. The authors identified only 11 trees (of 1138) with DBH (diameter at breast height) >40 cm that were the likely survivors of deforestation. Genetic diversity increased in younger generations and genetic differentiation decreased, and the authors suggest that wind pollination over 70 years reduced the negative genetic consequences of severe forest loss and led to genetic recovery of the population.

While many oaks have suffered demographic declines and range reductions due to habitat loss, fragmentation, and climate change, some species are expanding their ranges polewards in response to warming climates. In a newly established population of *Q. ilex* growing more than 30 km ahead of other stands in southwestern France, Hampe et al. [21] report a remarkably high fraction of immigrant genotypes (27%) in established offspring. These authors conclude that long-distance pollen flow can rapidly restore genetic diversity in leading-edge populations of oaks (and other wind-pollinated trees).

Studies have also used the microsatellite/paternity assignment approach to dig more deeply into *Quercus* male reproductive strategies. Lagache et al. [22] used a detailed paternity study and a spatially explicit individual-based mating model to investigate male fecundity in a mixed stand of *Q. robur* and *Q. petraea*. They report different male reproductive strategies for each species; *Q. petraea* trees dispersed their pollen over shorter distances and received less immigrant pollen than *Q. robur*. *Quercus rober* had a male reproductive strategy that favored pollen dispersal, having average pollen dispersal distances that were 40% greater than *Q. petraea*. There were also differences in interspecific sexual barriers, with *Q. petraea* having much stronger mating incompatibility. These results on differences in male reproductive strategies corresponded well to other evidence of these two species having different ecological strategies overall.

Rather than honing in on fine-scale details of pollination patterns at one site, Gerber et al. [23] chose to expand studies of pollination patterns across a much broader spatial scale. These authors studied gene flow (both pollen and seed dispersal) in eight mixed stands of white oak distributed across Europe, ranging from Spain to Sweden and Italy to Great Britain. The stands were mostly *Q. petraea* and *Q. robur*, but also included *Q. pubescens* and *Q. faginea*. The results of parentage analysis were used to simulate pollen (and seed) dispersal kernels using maximum likelihood. The results confirm that wind-borne pollen from outside the stand was high, 60% on average, but that it varied greatly among sites, ranging from 81% in Spain to 21% in the Denmark stand. Interestingly, pollen immigration rates were not correlated with stand size but estimated mean pollen dispersal distances were, ranging from 16 m in a small stand to almost 5400 m in a large stand. Long-distance dispersal events of almost 10 km were reported between scattered groups of trees.

Taken together, these recent studies further advance the paradigm of long distance pollen movement in oaks and extremely fat-tailed pollen dispersal kernels for their wind-dispersed pollen. As a consequence, even physically isolated oaks and oak stands are generally not reproductively isolated. Habitat loss and fragmentation will not likely signal a genetic doomsday for oak populations, although ecological and demographic losses loom large. However, some recent studies have revealed limitations and exceptions to the rule of long distance pollination, identifying populations that are experiencing restricted gene flow. A study of relictual populations of *Q. robur* at the extreme southwest edge of its range in western Spain reported that pollen exchange between stands was rare (2.6%), and the median pollination distance was just 30.5 m [24]. The restricted pollen dispersal resulted in substantial levels of genetic differentiation among stands. These authors also documented a phenological effect, with late flowering stands receiving more immigrant pollen than early flowering stands. Another study is notable because of its focus on a neotropical oak, *Q. segoviensis* [25]. Some of the populations sampled at the southernmost extreme of the species range in Nicaragua showed reduced genetic variability and substantial genetic differentiation, indicating reproductive isolation over relatively small geographic distances. Another neotropical oak, *Q. oleoides*, at the southern periphery of its range also showed genetic differentiation among populations [26]. This study was noteworthy because the authors investigated the role of phenology, environment, and distance in explaining genetic structure. The authors report no evidence of phenological isolation or isolation-by-environment and conclude that limited pollen dispersal was the main contributor to population genetic structure. Further investigations of tropical oaks are needed to determine whether pollination distances are more restricted for these species than for temperate oaks, at least at the margins of species ranges. Interestingly, even species with extensive pollen movement may have isolated populations. Although long distance pollination has been est ablished for the California species *Q. lobata* [16], population and landscape analysis found that a topographic feature, likely a mountain range, appears to be a barrier to gene flow, isolating at least one population at the southern periphery of its range [27,28].

Twenty-five years after researchers began using genetic markers to track pollination in oaks, we have gained a fairly detailed view of these patterns. In most, but not all, cases, we find evidence of extensive gene flow via pollen, and the only evidence to date of reproductive isolation occurs at the extreme peripheries of species ranges. This suggests that habitat fragmentation in oaks will generally not lead to genetic declines such as loss of genetic diversity. Both lagging- and leading-edge populations will continue to experience genetic exchange with core populations. Reestablishment or range expansion will be more constrained by arrival of acorns than by subsequent gene flow via pollen.

It is important to bear in mind that pollination distance is only one component of pollination biology, and that pollen abundance and pollen diversity are two different phenomena. Just because an oak tree will tend to receive pollen from near and far, and produce genetically diverse acorns, that does not preclude the possibility of pollen limitation for oaks. The studies cited here only document successful pollinations, but do not address the issue of whether pollen limitation (and reduced acorn production) occurs in either large stands or remote populations. Indeed, a number of studies have found a connection between masting in oaks and population-wide pollen availability [29,30]. For *Q. lobata*, high spring temperatures result in more synchronous flowering, which reduces pollen limitation and leads to high acorn production [30–32]. If phenological synchrony drives acorn production and masting behavior globally, an important but underappreciated ecological consequence of climate change could be changes in seed production patterns.

Another issue worth mentioning is also related to phenology. Because of long-distance pollination, most oaks will not be reproductively isolated by distance from other conspecifics; however, they may be isolated by flowering phenology. Different flowering times will isolate trees (or any plants) even if they are side by side. Microhabitat differences, local adaptation, climate change, and other factors may increase or decrease flowering synchrony and, thus, patterns of gene flow in oaks.

Finally, although the paradigm of long distance pollination in oaks has been firmly established over the last 25 years, it raises many intriguing questions for future studies in wind-pollination. How is it possible that the stigma of an oak flower gets fertilized by a pollen grain that has traveled hundreds of meters, rather than by the abundant pollen produced by neighboring trees? Could there be an element of female choice [33]? Does pollen have a maturation process that provides a fertility advantage for pollen that has been airborne for some time? Future oak research may reveal intriguing processes related to pollination, including phenological drivers, wind column dynamics, pollen properties, and mate choice. The next 25 years should bring exciting new insights.

Funding: This research received no external funding.

Conflicts of Interest: The authors declare no conflict of interest.

References

1. Ashley, M.V.; Dow, B.D. The use of microsatellite ahbnalysis in population biology: Background, methods and potential applications. *EXS* **1994**, *69*, 185–201. [PubMed]
2. Chase, M.R.; Moller, C.; Kesseli, R.; Bawa, K.S. Distant gene flow in tropical trees. *Nature* **1996**, *383*, 398–399. [CrossRef]
3. Dow, B.D.; Ashley, M.V. Microsatellite analysis of seed dispersal and parentage of saplings in bur oak. *Quercus macrocarpa. Mol. Ecol.* **1996**, *5*, 615–627. [CrossRef]
4. Dow, B.D.; Ashley, M.V. High levels of gene flow in bur oak revealed by paternity analysis using microsatellites. *J. Hered.* **1998**, *89*. [CrossRef]
5. Ashley, M.V. Plant parentage, pollination, and dispersal: How DNA microsatellites have altered the landscape. *CRC. Crit. Rev. Plant Sci.* **2010**, *29*, 148–161. [CrossRef]
6. Streiff, R.; Ducousso, A.; Lexer, C.; Steinkellner, H.; Gloessl, J.; Kremer, A. Pollen dispersal inferred from paternity analysis in a mixed oak stand of *Quercus robur* L. and *Q. petraea* (Matt.) Liebl. *Mol. Ecol.* **1999**, *8*, 831–841. [CrossRef]
7. Dow, B.D.; Ashley, M.V.; Howe, H.F. Characterization of highly variable (GA/CT) n microsatellites in the bur oak, *Quercus macrocarpa. Theor. Appl. Genet.* **1995**, *91*. [CrossRef]
8. Dow, B.D.B.D.; Ashley, M.V. Factors influencing male mating success in bur oak, *Quercus macrocarpa. New For.* **1998**, *15*, 161–180. [CrossRef]
9. Faegri, K.; van der Pijl, L. *The Principles of Pollination Ecology;* Pergamon Press: Oxford, UK, 1979; ISBN 9781483293035.
10. Okubo, A.; Levin, S.A. A theoretical framework for data analysis of wind dispersal of seeds and pollen. *Ecology* **1989**, *70*, 329–338. [CrossRef]
11. Craft, K.J.; Ashley, M.V. Pollen-mediated gene flow in isolated and continuous stands of bur oak, *Quercus macrocarpa* (Fagaceae). *Am. J. Bot.* **2010**, *97*, 1999–2006. [CrossRef] [PubMed]
12. Craft, K.J.; Ashley, M.V. Landscape genetic structure of bur oak (*Quercus macrocarpa*) savannas in Illinois. *For. Ecol. Manag.* **2007**, *239*, 13–20. [CrossRef]
13. Valbuena-Carabāa, M.; González-Martínez, S.C.; Sork, V.L.; Collada, C.; Soto, A.; Goicoechea, P.G.; Gil, L. Gene flow and hybridisation in a mixed oak forest (*Quercus pyrenaica* Willd. and *Quercus petraea* (Matts.) Liebl.) in central Spain. *Heredity* **2005**, *95*, 457–465. [CrossRef] [PubMed]
14. Nakanishi, A.; Tomaru, N.; Yoshimaru, H.; Manabe, T.; Yamamoto, S. Effects of seed- and pollen-mediated gene dispersal on genetic structure among *Quercus salicina* saplings. *Heredity* **2009**, *102*, 182–189. [CrossRef]
15. Nakanishi, A.; Tomaru, N.; Yoshimaru, H.; Kawahara, T.; Manabe, T.; Yamamoto, S. Patterns of pollen flow and genetic differentiation among pollen pools in *Quercus salicina* in a warm temperate old-growth evergreen broad-leaved forest. *Silvae Genet.* **2004**, *53*, 258–264. [CrossRef]
16. Abraham, S.T.; Zaya, D.N.; Koenig, W.D.; Ashley, M.V. Interspecific and intraspecific pollination patterns of valley oak, *Quercus lobata*, in a mixed stand in Coastal Central California. *Int. J. Plant Sci.* **2011**, *172*. [CrossRef]
17. Buschbom, J.; Yanbaev, Y.; Degen, B. Efficient long-distance gene glow into an isolated relict oak stand. *J. Hered.* **2011**, *102*, 464–472. [CrossRef]
18. Ortego, J.; Bonal, R.; Muñoz, A.; Aparicio, J.M. Extensive pollen immigration and no evidence of disrupted mating patterns or reproduction in a highly fragmented holm oak stand. *J. Plant Ecol.* **2014**, *7*, 384–395. [CrossRef]
19. Oyama, K.; Herrera-Arroyo, M.L.; Rocha-Ramírez, V.; Benítez-Malvido, J.; Ruiz-Sánchez, E.; González-Rodríguez, A. Gene flow interruption in a recently human-modified landscape: The value of isolated trees for the maintenance of genetic diversity in a Mexican endemic red oak. *For. Ecol. Manag.* **2017**, *390*, 27–35. [CrossRef]
20. Zeng, X.; Fischer, G.A. Wind pollination over 70 years reduces the negative genetic effects of severe forest fragmentation in the tropical oak *Quercus bambusifolia. Heredity* **2020**, *124*, 156–169. [CrossRef]
21. Hampe, A.; Pemonge, M.-H.; Petit, R.J. Efficient mitigation of founder effects during the establishment of a leading-edge oak population. *Proc. R. Soc. B Biol. Sci.* **2013**, *280*, 20131070. [CrossRef]
22. Lagache, L.; Klein, E.K.; Ducousso, A.; Petit, R.J. Distinct male reproductive strategies in two closely related oak species. *Mol. Ecol.* **2014**, *23*, 4331–4343. [CrossRef] [PubMed]
23. Gerber, S.; Chadœuf, J.; Gugerli, F.; Lascoux, M.; Buiteveld, J.; Cottrell, J.; Dounavi, A.; Fineschi, S.; Forrest, L.L.; Fogelqvist, J.; et al. High rates of gene flow by pollen and seed in oak populations across Europe. *PLoS ONE* **2014**, *9*, e85130. [CrossRef]
24. Moracho, E.; Moreno, G.; Jordano, P.; Hampe, A. Unusually limited pollen dispersal and connectivity of Pedunculate oak (*Quercus robur*) refugial populations at the species' southern range margin. *Mol. Ecol.* **2016**, *25*, 3319–3331. [CrossRef]
25. Ortego, J.; Bonal, R.; Muñoz, A.; Espelta, J.M. Living on the edge: The role of geography and environment in structuring genetic variation in the southernmost populations of a tropical oak. *Plant Biol.* **2015**, *17*, 676–683. [CrossRef]

26. Deacon, N.J.; Cavender-Bares, J. Limited pollen dispersal contributes to population genetic structure but not local adaptation in *Quercus oleoides* Forests of Costa Rica. *PLoS ONE* **2015**, *10*, e0138783. [CrossRef] [PubMed]
27. Gharehaghaji, M.; Minor, E.S.; Ashley, M.V.; Abraham, S.T.; Koenig, W.D. Effects of landscape features on gene flow of valley oaks (*Quercus lobata*). *Plant Ecol.* **2017**, *218*, 487–499. [CrossRef]
28. Ashley, M.V.; Abraham, S.T.; Backs, J.R.; Koenig, W.D. Landscape genetics and population structure in Valley Oak (*Quercus lobata* Nee). *Am. J. Bot.* **2015**, *102*, 2124–2131. [CrossRef] [PubMed]
29. Koenig, W.D.; Ashley, M.V. Is pollen limited? The answer is blowin' in the wind. *Trends Ecol. Evol.* **2003**, *18*, S0169–S5347. [CrossRef]
30. Pesendorfer, M.B.; Koenig, W.D.; Pearse, I.S.; Knops, J.M.H.; Funk, K.A. Individual resource limitation combined with population-wide pollen availability drives masting in the valley oak (*Quercus lobata*). *J. Ecol.* **2016**, *104*, 637–645. [CrossRef]
31. Pearse, I.S.; Koenig, W.D.; Knops, J.M.H. Cues versus proximate drivers: Testing the mechanism behind masting behavior. *Oikos* **2014**, *123*, 179–184. [CrossRef]
32. Koenig, W.D.; Knops, J.M.H.; Carmen, W.J.; Pearse, I.S. What drives masting? The phenological synchrony hypothesis. *Ecology* **2015**, *96*, 184–192. [CrossRef] [PubMed]
33. Craft, K.J.; Brown, J.S.; Golubski, A.J.; Ashley, M.V. A model for polyandry in oaks via female choice: A rigged lottery. *Evol. Ecol. Res.* **2009**, *11*, 471–481.

MDPI

St. Alban-Anlage 66

4052 Basel

Switzerland

Tel. +41 61 683 77 34

Fax +41 61 302 89 18

www.mdpi.com

Forests Editorial Office

E-mail: forests@mdpi.com

www.mdpi.com/journal/forests